PAPPUS OF ALEXANDRIA
AND THE MATHEMATICS OF LATE ANTIQUITY

This book is at once an analytical study of one of the most important mathematical texts of antiquity, the *Mathematical Collection* of the fourth-century AD mathematician Pappus of Alexandria, and an examination of the work's wider cultural setting. An important first chapter looks at the mathematicians of the period and how mathematics was perceived by people at large. The central chapters of the book analyze sections of the *Collection*, identifying features typical of Pappus' mathematical practice. The final chapter draws together the various threads and presents a fuller description of Pappus' mathematical agenda. This is one of very few books to deal extensively with the mathematics of late antiquity. It sees Pappus' text as part of a wider context and relates it to other contemporary cultural practices, opening new avenues to research into the public understanding of mathematics and mathematical disciplines in antiquity.

SERAFINA CUOMO is currently a Junior Research Fellow at Christ's College, University of Cambridge. She has held a post-doctoral Research Fellowship at the Max-Planck-Institut für Wissenschafts-Geschichte in Berlin and was for a year Assistant Professor in the Department of History at Iowa State University.

CAMBRIDGE CLASSICAL STUDIES

General Editors
P. E. EASTERLING, M. K. HOPKINS,
M. D. REEVE, A. M. SNODGRASS, G. STRIKER

PAPPUS OF ALEXANDRIA
AND THE MATHEMATICS OF LATE ANTIQUITY

S. CUOMO

CAMBRIDGE
UNIVERSITY PRESS

CAMBRIDGE UNIVERSITY PRESS
Cambridge, New York, Melbourne, Madrid, Cape Town, Singapore, São Paulo

Cambridge University Press
The Edinburgh Building, Cambridge CB2 8RU, UK

Published in the United States of America by Cambridge University Press, New York

www.cambridge.org
Information on this title: www.cambridge.org/9780521642118

First published 2000
This digitally printed version 2007

Chapter 2 of this book is a revised version of a chapter originally published in
Gattungen wissenschaftlicher Literatur in der Antike, ed. by Wolfgang Kullmann,
Jochen Althoff, Markus Asper. Tübingen: Narr, 1998 (ScriptOralia, Vol. 95).
© 1998 Gunter Narr Verlag Tübingen.

A catalogue record for this publication is available from the British Library

Library of Congress Cataloguing in Publication data
Cuomo, S. (Serafina)
Pappus of Alexandria and the mathematics of late antiquity / S. Cuomo.
p. cm. – (Cambridge classical studies)
Includes bibliographical references and index.
ISBN 0 521 64211 6
1. Mathematics, Greek. 2. Pappus, of Alexandria. Mathematical collections.
I. Title. II. Series.
QA22.C86   2000
510′.938–dc21   99-12317 CIP

ISBN 978-0-521-64211-8 hardback
ISBN 978-0-521-03689-4 paperback

CONTENTS

CONTENTS

ACKNOWLEDGEMENTS

Writing a book is one of the few opportunities you have to make a public profession of gratitude to anyone who has ever done anything for you – from your primary school mathematics teacher who told you to think about what you were doing, to your secondary school literature teacher who told you that you could do better, to your extended family who kept wondering why you could not get a proper job like anybody else. But I'll try to be brief. I would like to thank first of all my friends. Thanks also to Richard Ashcroft, Domenico Bertoloni Meli, David Fowler, Marina Frasca Spada, Ofer Gal, Peter Garnsey, Nick Jardine, David Sedley, Skuli Sigurdsson, Liba Taub, the anonymous referees for the press, the audiences of various conferences and seminars where I have presented parts of the book, and especially Reviel Netz, for comments and criticisms. A special thank you to my many teachers, especially Professor Giovanni Casertano, and very very special thanks to my thesis supervisor, Professor Geoffrey Lloyd. He has been the best of supervisors and the kindest of men. I would like to acknowledge the financial support of Christ's College, Cambridge and the Max-Planck-Institut für Wissenschaftsgeschichte in Berlin. I would also like to thank the editors of Cambridge Classical Studies for accepting the book for publication. Finally, I dedicate this book, such as it is, to my husband with love.

INTRODUCTION

The arch of Constantine in Rome was probably begun in AD 312, to celebrate the victory of the Emperor over Maxentius at the battle of the Milvian Bridge. The monument can be defined "a true palimpsest of Roman Imperial sculpture." Four of the bas-reliefs that adorn the arch were taken from a monument built in Trajan's time; several others were originally found in another arch, for Marcus Aurelius; some of the columns and other architectural elements date from Domitian's time. The only sculptures which were made in Constantine's time are two medallions and the friezes.[1]

The mixture of past and present in the arch is not limited to the borrowing of pieces from different epochs: the heads of the original emperors in the bas-reliefs (Trajan and Marcus Aurelius) have been cut off and replaced with heads of Constantine.

The art historian Bernard Berenson had this to say about the arch:

A sinister fact meets us at the start. It is that the charming hunting scenes in the medallions and the military displays inside the arches ... were not made for this edifice but taken from structures dedicated to Trajan, Hadrian and Marcus Aurelius. This curious procedure may have been due not only to hurry to get things ready in time but to a feeling that nothing could be done there and then as worthy of the occasion. Explain it as you will, it was a confession of inferiority to the past, whether economic or artistic.[2]

More recently, on the other hand, the attempt has been made to

look more closely at how the *spolia* [the pieces from earlier periods] are incorporated into the Arch, and particularly at how they are made to interact with the Constantinian decoration. An examination of the concordances

---

[1] This information from Lugli (1970), 376–378. [2] Berenson (1954), 13–14.

between the *spolia* and the Constantinian reliefs reveals the sophisticated way in which the past is presented as a referent for the present, and, by implication, the future.[3]

The *Mathematical Collection* of Pappus of Alexandria was written some time around AD 320. It is a collection in more than one sense: its various parts were probably composed separately and put together at a later date, and they use material from different epochs.[4] Pappus draws on a variety of sources, ranging from Euclid (early third century BC) to Sporus (late third century AD).

This much has been said about Pappus and the *Collection*:

In the later Hellenistic period ..., the main stream of Greek mathematics ... experienced a deep and permanent decline.... Pappus of Alexandria is [an] author in this degenerate tradition .... The period around the fourth century AD has often been described as a "Silver Age" of mathematics, an illusion for which the bulk of Pappus's extant work, and the abundance of information uniquely preserved in it, are largely responsible. In fact the few occasions on which Pappus claims something as his original discovery give little evidence of a fertile mind. Nevertheless his reputation ... was high – deservedly so, according to the debased standards of his time.

Also,

Pappus knew that he lived in a period of decline, and his reverence for the "ancients" is matched by his disdain for his contemporaries.[5]

I do not think that this is a good description. It seems to me that ancient mathematics to some extent, and late ancient mathematics to a massive degree, suffers from what we could call, borrowing A. I. Sabra's words about the historiography of Islamic science, "the marginality thesis." That is, the belief that

scientific ... activity ... had no significant impact on the social, economic, educational and religious institutions; that this activity remained itself un-

---

[3] Peirce (1989), 411.

[4] For the hypothesis that the *Collection* in its present form was put together at a later stage cf. Ziegler (1896), cols. 1084–1106; Jones (1986a), I 15 and 24–25. There is a passage that refers to "this, the third book of the collection" (30.21–22; henceforth all references without further specification are to the *Collection*), but Jones thinks this "is not significant: he or his redactor would automatically have changed such a phrase as 'in this letter' to one more appropriate for inclusion in a volume of collected works" (1986a), I 17.

[5] Both by Alexander Jones in (1986a), I 1 and (1994b), 65, respectively.

affected by these institutions, except when it was finally crushed by their antagonism or indifference; and that those [engaged in this activity] ... constituted a small group of scholars ... whose work and interests were marginal to the central concerns of ... society.[6]

This belief is especially strong when the scientific activity in question is mathematics.

Again in a parallel with art, there is a sense in which the *Venus* of Milo is "objectively" beautiful and the statement that two and two makes four is "objectively" true. The precise nature of this sense is in either case a topic for the philosopher, and I will not tackle it in this book. What concerns me is that, on the basis of this sense, not only do historians of mathematics feel entitled to pass judgment on previous mathematical texts, because there are universal standards against which their adequacy can be assessed; they also feel entitled to abstract from, or simply ignore, the historical circumstances in which a mathematical text is produced – we are to believe that it makes more historical sense to see Pappus within a mathematical framework whose characteristics were (allegedly) fixed in stone by Euclid and Archimedes, than it does to inscribe his activities within a fourth-century AD framework.

This book is an attempt to produce a historical analysis of the *Mathematical Collection* of Pappus of Alexandria (given its size, only a partial analysis), and at the same time to explore its wider cultural contexts. Ultimately, I hope to show that the mathematics of late antiquity deserves a place both in the history of science and in the history of antiquity.

In the first chapter I will look at the wider picture: who were the mathematicians in and around the fourth century AD, and how was mathematics perceived by people at large? As a result of my (once again, necessarily partial) enquiry, I will show that mathematics, in its various facets, was more common and more familiar than is usually thought; that there was some notion of mathematical "professions"; that an important part of the self-image of those professions had to do both

---

[6] Sabra (1987), 229. See also for new perspectives on the history of mathematics Lloyd (1994); Goldstein (1996); Goldstein, Gray & Ritter (1995); Høyrup (1996a) and (1996b); Vitrac (1996).

with expertise and with ethical qualities; that reflections on mathematics as a form of knowledge were often accompanied by boundary-drawing operations, aimed at establishing differences between "good" and "bad" mathematicians, and between mathematicians and other people. The historiography of the last fifteen years has established that reference to the past and to tradition was a fundamental element in the cultural practices of late antiquity. I will argue that this was true also, perhaps unsurprisingly, of mathematical practices.

Chapter two will closely examine the fifth book of Pappus' *Collection*. I will argue that both the structure of the mathematical argumentation and a number of explicit "second order" statements show that Pappus is addressing a non-specialized public; that he wants to convey the importance of mathematical knowledge and at the same time the importance of himself as repository of that mathematical knowledge; that he does so by contrasting, on the one hand, his type of knowledge to that of some unidentified philosophers, and human mathematical knowledge to that of bees. I will also start to pin down some features typical of Pappus' mathematical practice: an interest in the criteria of validity for the solution to a problem; an interest in generalizations and particularizations (sub-cases) of mathematical statements; an "opportunistic" attitude to his sources, which are quoted, modified or acknowledged according to the aims of a particular book or a particular passage within a book.

This initial picture will be further substantiated and complicated by evidence from the last book of the *Collection*, where Pappus gives a long definition of mechanics, explores some issues related to it and argues throughout for the complementarity of mathematics and mechanics (chapter three of the present volume), and from Pappus' account of the duplication of the cube and of "linear" curves (chapter four of the present volume). The several threads are drawn together in chapter five, where I will give a fuller description of what I take to be Pappus' mathematical agenda, and will show that this latter presupposes that past traditions were constructed to give authority and prestige to present practices and present

4

practitioners. Moreover, I will argue that, once again in line with the cultural phenomenology of his time, Pappus assembles an image of the "good" mathematician as someone who possesses both expertise and ethical virtues.

Now, I do not wish to offend my learned reader (especially not at such an early stage), but the next question may well be "Pappus who?" We do not know much about Pappus of Alexandria. His main work, the *Mathematical Collection*, consists of eight books, one and a half of which are lost, and deals with topics ranging from astronomy and mechanics to arithmetic and "classical" geometry. He also wrote commentaries on Ptolemy's *Almagest* and on Euclid's *Elements* and a *Geographia*. An enchantment oath mentions a "Pappus the philosopher" as its author, but identification is dubious. Of other various works attributed to Pappus only the titles have survived.[7]

That Pappus of Alexandria did live in Alexandria is not entirely certain. The toponymic is attached to the name in the (probably later) titles of his surviving books and in the relative entry in the tenth-century *Suda* Lexicon, but it might be that Pappus' family hailed from the Egyptian city, while he was born, or spent most of his life, abroad.[8] That said, it is quite likely that Pappus lived in Alexandria. He seems to have had access to a good number of books and refers to several contemporaries: philosophers, mathematicians and astronomers. All this points to activity in a large centre rather than a

[7] Of the commentary on the *Almagest*, only the parts relative to books 5 and 6 are extant. Of the commentary on Euclid, the part relative to book 10 survives in an eleventh-century Arabic translation. The *Geographia* is extant in a much revised Armenian version, an English translation of which is in Hewsen (1971). The oath is in *Collection des anciens alchimistes grecs* I.12; see Tannery (1896). Works of which we have only the titles are: *Rivers of Libya* and *Interpretation of dreams*, listed in the *Suidae Lexicon* P265; a commentary on Ptolemy's *Planispherum*, mentioned by the *Fihrist* of ibn al-Nadim (I have this from Jones (1986a), I 12); a commentary on Diodorus' *Analemma*, mentioned by Pappus himself, 246.1; an astrological almanac attributed to him by an eleventh-century manuscript (again in Jones (1986a), I 13, who also quotes a fragment from a thirteenth-century astrological manuscript where Pappus is indicated as the source for an anecdote).

[8] E.g., the parents of Proclus "Lycius" were natives of Lycia, but he was born at Byzantium and lived between Alexandria and Athens. Or again Eratosthenes of Cyrene spent most of his life in Alexandria.

small one. Also, an inscription dating from the second or early third century AD has been found in Alexandria, with which Basso, son of Strato, administrator of the temple of Serapis, dedicated a statue to Pappos Theognostos "for the good."[9]

Determining a precise date for Pappus is problematic. According to the *Suda*, he lived under the Emperor Theodosius (AD 379–395), whereas a scholion to a Leiden manuscript of Theon's edition of Ptolemy's *Handy Tables* places him under Diocletian's reign (AD 284–305, but the scholion has the dates 284–308).[10] Pappus himself, in his commentary on Ptolemy, mentions as quite recent a sun eclipse which A. Rome first identified with one that took place in AD 320.[11] Considering that at least some parts of the *Collection* were composed after the commentary,[12] the indication that Pappus lived under Theodosius is entirely likely – he may even have been born towards the end of Diocletian's reign and have lived to a very ripe old age. I will assume that he spanned the best part of the fourth century AD.

As we will see in the course of the volume, Pappus is mentioned by several later Greek authors, Proclus, Marinus, and Eutocius, and he was known by several Islamic writers.[13] The question of whether his works were known in the medieval West is undecided: the main manuscript of the *Collection* (*Vat. gr.* 218, dated to the tenth century) seems to have been in the Vatican library by at least 1311 and possibly already by 1266, but it does not seem to have been copied until much later. The issue has been complicated by the fact that some

---

[9] The name Pappus may denote Jewish origins, see Netz (forthcoming a); the inscription is number 44 in Kayser (1994). The enchantment oath mentioned above (see n.7) invokes a Christian God, his cherubs and perhaps the Trinity (the God is described as one in form but not in number); authorship of it could denote Christian or possibly Manichean affiliations.

[10] Bulmer-Thomas (1974), 294.

[11] *In Ptolomaei Almagestum* 180.8–181.23, see X–XIII.

[12] The commentary is mentioned in book 8 at 1106.13–14: "and as it is proved in the scholion to the first of the mathematical [books?]" (Καὶ ὡς ἐν τῷ εἰς τὸ πρῶτον τῶν μαθηματικῶν σχολίῳ δέδεικται). Here and henceforth all translations of the *Collection* are mine, unless otherwise stated and unless they are from book 7. Translations from book 7 of the *Collection* and from any other book are as from bibliography unless otherwise stated.

[13] For the Islamic tradition see Jones (1986a).

propositions contained in the Polish natural philosopher Witelo's *Perspectiva* (*ca.* 1270) bear a very close resemblance to some results of book six of the *Collection*.[14] Moreover, in his *On the five regular bodies* (1480) the Italian painter Piero della Francesca describes the construction of some semi-regular solids which are otherwise only known through book five of the *Collection*.[15]

The first printed edition of the *Collection* (in Latin translation) came out in Pesaro (Italy) in 1588 and was based on work left unfinished by Federico Commandino;[16] it did not include book two, which was first published by John Wallis in Oxford in 1688. The early modern period saw a surge of popularity for Pappus: René Descartes took one of the problems of book 7 of the *Collection* as the starting point for his 1637 *Géométrie*[17] and Guidobaldo dal Monte (1545–1607) declared:

[N]o one could, I believe, blame me for following [Pappus] as my leader.... I wish that the ravages of time had not caused any loss in the writings of so great a man. For such a thick mist of ignorance would not have covered almost all the earth, nor would there have been such ignorance of the subject of mechanics.[18]

Friedrich Hultsch's standard edition, published in 1875, stands in need of revision, as Alexander Jones has abundantly shown with his own new edition of book 7. In particular, the German philologist was what one could call interpolation-happy – as we will see on several occasions, he often bracketed

---

[14] Unguru (1974). For the manuscript tradition see also Rome (1938) and (1948); Treweek (1957); Solomon (1983); Jones (1986b).

[15] See Clagett (1964). Jones (1986a), I 50 thinks that "an independent discovery is very probable."

[16] On the complicated history of the 1588 edition, see Passalacqua (1994).

[17] Descartes (1637), chapter 1. The problem in book 7, 678.15–24: "If three straight lines are given in position, and from some single point straight lines are drawn on to the three at given angles, and the ratio of the rectangle contained by two of the (lines) drawn onto (them) to the square of the remaining one is given, the point will touch a solid locus given in position, that is, one of the three conic curves. And if (straight lines) are drawn at given angles onto four straight lines given in position, and the ratio of the (rectangle contained) by two of the (lines) that were drawn to the (rectangle contained) by the other two that were drawn is given, likewise the point will touch a section of a cone given in position."

[18] Dal Monte, *Mechanicorum liber* (1577), Introd., in Drabkin & Drake (1969), 244.

out entire paragraphs on dubious criteria of "inner consistency" or "redundancy."[19] The *Collection* has been translated into French and (partially) into German (books 7 and 8) and, more recently, English (book 7).[20] On the whole, very little work has been done on Pappus, or indeed on the mathematics of late antiquity.[21] I will discuss previous interpretations as they become relevant in the course of the next chapters.

Now let us start.

---

[19] Jones (1986a), I 18–20, 65; Knorr (1989), 8n7, who quotes J. L. Heiberg's comment that: [Hultsch employs] "the knife where the harm could have been healed by the milder cure of emendation".

[20] French trans. in Ver Eecke (1933a); English trans. in Jones (1986a), which also contains details of C. I. Gerhardt's 1871 German translation.

[21] Many of the books on ancient mathematics devote only few pages to either: see Cantor (1907); Boyer (1939); van der Waerden (1954); Clagett (1956); Eves (1976); Edwards (1979); Gericke (1992).

CHAPTER I

THE OUTSIDE WORLD

The only ancient mathematicians whose lives we know anything about are those who died in spectacular and gruesome
circumstances: Archimedes, slain in 212 BC by Roman soldiers
on the rampage, Syracuse having been taken after more than
two years of siege; Hypatia, torn to bits by a lynching Christian mob in the streets of Alexandria in AD 415. Pappus,
however, as well as Euclid, Apollonius or Ptolemy, must have
had a quiet life, so he remains just a name to us. And he is one
of the "famous" mathematicians, as distinct from the relatively substantial number of anonymous people whose lives
and professions had something to do with mathematics, who
studied it or had studied it, who used it cursorily for their job
or taught it at various levels.

My aim in this chapter will be to gather some evidence
about the mathematical practices of late antiquity *apart* from
what we find in the books of the famous mathematicians – we
will take a look at the outside world.[1] This will imply looking
at a variety of sources which are not usually considered relevant for the history of mathematics. Why not? It seems
somewhat perverse to dismiss as irrelevant evidence taken
from astrology, land-surveying, architecture, and mechanics
because all those people "weren't mathematicians really."
Astrologers called themselves *mathematici* and land-surveyors
geometres and their writings are replete with calculations,
praises of mathematics and statements that mathematical
knowledge was an essential part of their activities – yet such
claims are discounted as mere rhetoric. Since those people

---

[1] Of course, given my limitations of time, space and expertise, this analysis will be
partial – just scratching the surface of what could be termed "public understanding" of mathematics in antiquity.

only actually used mathematics in order to carry out very simple calculations or apply elementary geometrical theorems, they should not belong in the same story as Euclid or Archimedes or, for that matter, Pappus.

I will start from the assumption that it is irrelevant for my purpose whether the mathematics they employed was simple or complex, or even whether they actually made use of mathematics at all. I am interested above all in whether they saw their work and their type of knowledge as having to do with mathematics and whether they constructed their self-image accordingly, especially when it came to contrasting their form of knowledge with other forms of knowledge. If claiming affiliation with mathematics was a rhetorical move, then we should ask ourselves how that rhetoric came into place, and why the people who were deploying it thought that their rhetoric could work.

What was the nature of the relations between "those people" and people like Pappus? When we look at the *Collection* in more detail, we will see that the presence of other people who claimed affiliation with mathematics did have an impact on it. Apart from that, I have no ready answer. I do not suggest causal links; rather, I intend to provide some background to what Pappus was doing, in the hope that the reader will share my sense that when the *Collection* was being written there was a world out there where ideas about mathematics and mathematicians loomed larger than is usually thought – a world that should not be ignored.

1.1 Mathematics and the stars

That astrological texts could be fruitfully used by ancient historians was an idea first put forward by Lynn Thorndike, revived by Ramsay MacMullen and lately applied with interesting results by Tamsyn Barton.[2] Part of the operation consists in seeing how possible star-determined destinies were

---

[2] Thorndike (1913); MacMullen (1971); Barton (1994a). See also Cumont (1937), 87 ff.

valued on the desirability scale, what was said about social ascent, reversals of fortune, sexual mores, and so on. I will be looking at how mathematical professions fared as a possible destiny – which ones they were, with what other professions they were grouped and whether they were generally seen as a bad or good thing.

Our principal source will be a Sicilian writer, Julius Firmicus Maternus, whose main work, *Mathesis*, was written around AD 337 in Latin for Mavortius, a government official, with the declared aim of providing the non-expert with a translation of useful bits from Egyptian and Babylonian astrological texts.[3] For his theories and methods of interpretation he draws upon previous astrologers, especially Dorotheus Sidonius (AD 25–75) and Vettius Valens (second century AD), but his frequent references to contemporary official titles and posts testify that his evaluation of such aspects is updated.[4] I have chosen to focus on Firmicus Maternus because he devotes a great deal of attention to jobs and professions, but I will use evidence from other late fourth-century astrological treatises as well: Paul of Alexandria's *Elementa Apotelesmatica* (*Elements of astrology*) and Hephaestio's *Apotelesmatica*.

Firmicus Maternus is a good example of the care ancient astrologers devoted to their self-image: his work contains a section about the opponents of astrology and an outline of the profession (*Qualis vita et quale institutum esse debet mathematicis*) which includes a whole code of behaviour for practitioners.[5] As well as possessing generic moral and social virtues, they must be close to the divinity, be available to the public, turn down pecuniary rewards and avoid trouble by refusing to make predictions about the Emperor.[6] A bound-

---

[3] The significance of Firmicus being Sicilian is that he was fluent in Greek and felt particularly close to the Hellenic heritage: Archimedes is by him called "my fellow citizen from Syracuse," *Mathesis* II 148.23 (II refers to the second volume of the Teubner edition; I have modified the translations in parts). Mavortius is mentioned by Ammianus Marcellinus, *Res Gestae* 16.8.5; cf. Rhys Bram (1975) for epigraphical evidence about him.

[4] MacMullen (1971), 221.

[5] *Mathesis* 4.9 ff. and 85.5 ff. respectively.

[6] *Math.* 85.3–89.5. Cf. Barton (1994a).

ary is drawn between bad astrologers and good ones: those latter are distinguished by their ethical qualities, as well as by the extent of their knowledge. Incompetent astrologers bring the discipline into disrepute: Maternus contrasts the *prudentissimus mathematicus* with the inept one both in terms of expertise (as against "fallacious and heedless ... ignorance") and of honesty in pursuing one's art.[7] Astrology is equated to "philosophy and divine knowledge"[8] and, because of its certainty, he claims, it compares favourably with the disputes and inconsistencies of the very people who attack it – for instance, the enemies of astrology have never been able to agree among themselves about the nature of the gods, and even Plato and Aristotle are at variance about the essence of the soul.[9]

The astrologer's expertise includes mathematical knowledge – the question of *what* mathematical knowledge is problematic. Most of the actual horoscopes on papyrus or ostrakon that have come down to us (ranging from the first century BC to the fourth AD, with the majority dating to the second or third AD) are not very detailed, and suggest that they were cast and interpreted by people who did not necessarily know much about the geometrical complexities of planetary motions and may have had a minimum of calculating skills. Many practitioners of astrology must have carried out their job with the help of planetary tables and instruments, for which there is quite extensive evidence, both material and textual.[10] On the

---

[7] *Math.* 10.15–21 and 11.2–11.

[8] *Math.* 4.14.

[9] *Math.* 4.29–5.31. Cf. Ptolemy, *Apotelesmatica* 2.7.20–8.1 about bad astrologers: "[Most] deceive the vulgar, because they are reputed to foretell many things, even those that cannot naturally be known beforehand, while to the more thoughtful they have thereby given occasion to pass equally unfavourable judgement upon the natural subjects of prophecy. Nor is this deservedly done; it is the same with philosophy – we need not abolish it because there are evident rascals among those that pretend to it." Hephaestion refers to the "truth" which can be reached in astrological observations: e.g. *Apotelesmatica* 81.17; 145.17. See also Hübner (1990).

[10] On horoscopes see Neugebauer & Van Hoesen (1959); Baccani (1992); Jones (1994a). On tables see Neugebauer (1975), section V A 2 and the more recent bibliography in Jones (1994a). On astronomical instruments the most recent is Turner (1994).

other hand, we have several more detailed horoscopes, where the number of data considered was greater and a substantial amount of knowledge, and possibly observations, were involved. It would seem that there was a range of competences which may have corresponded to different customers or different demands: some astrologers would have commanded a higher price for highly personalized, complex birth-charts, some others would have put together more ordinary horoscopes. We can also imagine that the same astrologer could have catered differently for different clients.[11]

Firmicus Maternus confirms this picture: he contrasts calculations with purely visual observations,[12] and distinguishes between the task of working out the motions of the heavenly bodies, which is more difficult and therefore only practicable for the expert practitioner who wants to make a very accurate prediction, and the mere *interpretation* of the effects of the motions, which is easier.[13] Also, he and Paul and Hephaestio all repeatedly insist on accuracy; they frequently refer to, or report, calculations and often draw up tables or mention instruments that can help achieve better results (the astrolabe, the gnomon, the so-called *sfaera barbarica*, which was probably a type of armillary sphere).[14] The type of mathematics involved may not have been very complex, but that need not interest us: what matters is that mathematics is presented as a

[11] For a similar scenario among medical practitioners, see Pearcy (1984); Lloyd (1993); Barton (1994b).

[12] *Math.* 69.11.

[13] *Math.* 13.29–15.6.

[14] In his commentary on Ptolemy's *Syntaxis* 5.1 and 5.12, Pappus describes in detail an astrolabe (3.11 ff.) and a parallactic instrument (70.10 ff.), and often mentions ἀκριβεῖα (mostly of data). Paulus Alexandrinus, *Elementa Apotelesmatica* 33.11–12; 79.10–11, uses Ptolemy's *Handy Tables* as a guarantee of the accuracy of some data. Hephaestio also mentions accuracy often: e.g. *Ap.* 84.31; 85.13–16; 94.26–95.7 (he recommends repeated experience and practice in order to reach greater accuracy and come closer to the truth); 133.1–2; 197.24–25; 327 23, and *Epitomae* 41.3; 216.22–23. The importance of accurate calculations is also stressed by Manilius, *Astronomica* e.g. III.218 ff. (early first century AD). As for instruments, there are relevant passages at Firmicus Maternus, *Math.* II 315.8 (gnomon); 278.15, II 174.25, II 284.1, II 288.14, II 294.12–359.11 (*sfaera barbarica*); Paulus Alexandrinus, *El. Ap.* 80.13, 20 (astrolabe); Hephaestio, *Ap.* 51.10–12; 88.24 (both about an astrolabe); 234.24 (in connection with a ὑδροσκοπίον, perhaps a water-clock) and *Ep.* 2.12; 269.10.

kind of knowledge that grants better results and a higher degree of expertise.

Let us see now how mathematical professions are represented in the treatise. Astrologers figure both as *astrologi* and as *mathematici*. Some distinction between the two terms probably existed (perhaps *astrologi* indicated a lower level of competence), but they generally share the same associations, apart from a couple of cases in which *mathematici* (without *astrologi*) are grouped with long-haired philosophers, or with those "who discover and learn by themselves what has not been handed down to them on the authority of others."[15] *Astrologi* are involved in a range of other activities: most often with *haruspices* and dream interpreters and, in the same crowd, with doctors or *archiatri* (court doctors),[16] but at times also with teachers and geometers, orators, "ingenious inventors" and grammarians and poets.[17]

On the whole, being an astrologer is a rather positive prospect, comparable to that of being a doctor or an orator. Indeed, such associations reinforce our knowledge of the links between astrology and medicine, and underline not only the predictive nature of both disciplines, but also the public persona of astrologers, the fact that casting and, above all, interpreting birth-charts implied some capacity for rhetorical performance.[18] Out of a total of fourteen occurrences in the

---

[15] *Math.* 157.17: "sacerdotes magos archiatros mathematicos et per se invenientes atque discentes, quicquid illis non est alieno ⟨traditum⟩ magisterio."

[16] *Math.* 162.3; 165.25 (both with doctors); 185.26; 252.12; II 25.8; II 335.6; II 340.21 (this latter on their own). Cf. the same association in Paulus, *El. Ap.* 64.6 (who puts together diviners, dream interpreters, astrologers, augurs and those who partake in mysteries) and Hephaestio, *Ap.* 147.4–7 (where a certain configuration of Venus and Mercury produces souls apt to investigate things that are hidden, i.e. magicians, meteorologers, mechanicians, astrologers, wonder-workers, augurs, dream-interpreters, philosophers) and 171.9 (magicians, astrologers, people who speak in precepts and make predictions).

[17] *Math.* 159.1; 263.18 and II 24.12; 263.25; II 24.12, respectively.

[18] For these themes in astrology and medicine, cf. Barton (1994b). Hephaestio, *Ap.* 146.13–21 has astrological souls in a group of souls inclined to politics, love of fame and divination, and at *Ap.* 3.7 says that the ancient Egyptians combined astronomy and medicine in making predictions (hence, the ἰατρομαθηματικοί already mentioned by Ptolemy *Ap.* 16.7 ff.).

Mathesis, doctors are associated with predictive professions (*haruspices*, astrologers) four times,[19] as well as once with *herbarii*,[20] once with dyers, perfume-makers, musicians and athletes[21] and once with orators and lawyers. Those latter, however, are the doctors who are "ennobled by their talents in their professions."[22] The two occurrences of *archiatrus*, a title which apparently denoted the top of the medical profession, are both in the company of astrologers and *haruspices*.[23]

Architects are mentioned three times, of which twice in connection with sculptors and builders;[24] mechanicians three times, of which once with sculptors, poets and musicians.[25] Several categories of technicians are present (*artifex, ex-inventor, fabricator, inventor, organarius*), but the differences between them are unclear. The destinies associated with them are as a rule quite favourable: some craftsmen will make a living out of their art and learn by themselves what is not handed down to them; others, more favoured by the stars, will be friends of kings; some public builders will get the greatest dignity from their job.[26]

Geometers are side by side with philosophers teachers of grammar, star-gazers, experts in *sacrae litterae*, orators and lawyers, or with astrologers and other diviners or, again, with scholarly grammarians, orators and teachers.[27] Some people's horoscopes indicate them as *tabularii* (accountants), tax experts[28] or discoverers: of calculations and instruments; of calculation, music, signs (*notae*) and difficult letters; together with orators, grammarians, doctors and musicians or with

---

[19] *Math.* 162.3; 165.25; 157.17; 263.18.
[20] *Math.* 168.8. Hephaestio, *Ap.* 170.23–29 has doctors, smiths, architects and measurers together.
[21] *Math.* 220.27.
[22] *Math.* II 338.11: "quos professionis suae nobilitet ingenium."
[23] *Math.* 157.17; 263.18. On *archiatri* see Nutton (1977).
[24] *Math.* II 349.1; II 331.21; II 322.18 (this latter has naval architects).
[25] *Math.* II 23.24 (with sculptors, etc.); 186.23; II 341.6–7 ("mechanicus qui instrumenta bellis faciat necessaria").
[26] *Math.* 162.15; II 344.12 and II 72.8, respectively.
[27] *Math.* 155.24; 159.1; 164.3, respectively. Hephaestio, *Ap.* 149.28–30 has philologists, geometers, mathematicians and unspecified "wise people" together.
[28] *Math.* II 307.4; II 75.15, respectively.

orators and teachers.[29] Of some people it is forecast that they will make a living out of reckoning and calculation,[30] while those who, more vaguely, "excel in numbers" are grouped together with public teachers, orators, grammarians and jurists.[31]

On the whole, a remarkable number of people had some involvement with mathematics written in the stars for them, and such a career would have constituted quite a desirable destiny. Also, jobs to do with numbers and calculations, often carried out for the general public, were usually included among the educated professions. In sum, activities associated with mathematics were highly visible, had a rather positive public profile and were generally linked to intellectual practices.[32]

Firmicus Maternus himself shows that within his discipline expertise and good, accurate practice were explicitly linked to the use of mathematics. Moreover (and we will see that astrology is not the only case) boundaries were drawn between good and bad practice on the basis both of ethical qualities and of the extent of a person's knowledge.

1.2 Mathematics in the workshop

Land-surveyors, architects, mechanicians, "technicians" of various sorts were all pretty visible in the scenario of late antiquity, as we noticed in looking at Firmicus' treatise and will continue to see in the section on laws. Sharp distinctions be-

---

[29] *Math.* 172.6; 161.3; 233.17; 170.21, respectively. There also people expert in calculation who interpret the course of the stars, 211.4 and 233.18. Hephaestio, *Ap.* 170.2–3 groups together lawyers, people in charge of offices – possibly of account offices (λογιστηρίων προϊσταμένους) – teachers and leaders of crowds.

[30] *Math.* 158.12 and 183.25: "computus aut calculus." Cf. also souls described as ἐπιλογιστικάς, together with, among others, souls which are high-minded and apt for science (μεγαλόφρονας and ἐπιστημονικάς), in Hephaestio, *Ap.* 146.24–26.

[31] *Math.* 182.2.

[32] Gregory of Nyssa (second half of fourth century AD; *Epistula* 17.24), wanting to make the point that the virtues of the master become the virtues of the pupils, declares that a smith will not become a weaver, nor will a weaver become a rhetorician or a geometer. Note that, although the two artisanal categories of smith and weaver are kept well separated, rhetor and geometer are taken to be somewhat alike.

tween one category and the other are hard to draw:[33] the architects for whom we have evidence include people who were
credited with ship-building and supervision of war-machines –
an architect could probably double as an engineer whenever
need required. Also, some terms are ambiguous: the Latin
geometres or *geometra*, whose primary meaning is "geometer," is often translated as "land-surveyor" unproblematically. A more balanced view would be to accept that the term
retained some ambiguity, thus reflecting the combination of
skills, "practical" and "theoretical," that the practitioners
themselves often possessed. It is sometimes clear from the
context that a land-surveyor is meant, sometimes a geometer;
at other times pinpointing one meaning in preference to the
other is simply not possible.[34]

Generally speaking, the social status of technicians in late
antiquity seems to have been upwardly mobile.[35] Architects,
for instance, received a great deal of recognition. Symmachus,
who was prefect of Rome between AD 384 and 391, had to
investigate a case involving two people who had senatorial
rank and had been trusted with public money in order to build
a basilica and a bridge: Auxentius and Cyriades, the latter a
"comes et mechanicae professor."[36] They were suspected of
embezzling funds, and pretty soon started to accuse each
other. The equal status of the two made a separate inquiry
necessary, and Symmachus decided to charge the master
craftsmen ("fabrilis artis magistros") with an assessment of
the situation as far as the works were concerned. Before the
results were produced, Auxentius fled town, and was replaced
by another senator, the notary Aphrodisius.

---

[33] Downey (1946–48), 109, argues that μηχανικός denotes someone superior to ἀρ-
χιτέκτων. Cf. also Coulton (1977); Bulmer-Thomas (1981); Mansuelli (1985);
Donderer (1996); Anderson (1997).

[34] Cf. Dilke (1971), 44; Schindel (1992), 377 and 377n18 for analogous views.

[35] Cf. Clarke (1971), 113; Cracco Ruggini (1971); Williams (1985), ch. 2; Bowman
(1992).

[36] Symmachus, *Relationes* 25–26; the case went on from AD 382 to 387. Auxentius
has been identified with the person celebrated in *IGR* 3.887, an inscription which
commemorates a bridge built on the river Sarus in Cilicia, see Grégoire (1927–8);
Vera (1981). Donderer (1996) has further examples of third- and fourth-century
architects with quite a high social status based in e.g. Antioch and Rome.

The people who had laid the foundations for the bridge were then interrogated, and it turned out that a part of the bridge, left unfinished (or defective, *incohata*), had been destroyed by the violence of the river, with the cost for repairs assessed by the *artifices* (a further step down the hierarchy at the building site) at twenty solids. Cyriades assured the authorities that repairs could be easily carried out, but further investigation revealed that in part of the bridge the stones were not fitting together. Cyriades then claimed that the job had initially been done properly according to his instructions ("consilio suo et ratione artis"), but that Auxentius had later deliberately spoiled the work, filling the gaps with straw and weeds, in order to make him (Cyriades) look incompetent. A diver ("urinandi artifex"), however, testified that the use of straw and weeds was not aimed at discrediting Cyriades, but at ensuring stability – it was a matter of using a different technique. Finally, the witness whose testimony was at variance with the others (it is not clear who the poor fellow was – maybe the diver) was subjected to torture and confessed that Cyriades had threatened him, but he was not believed because it was thought that the confession was made simply to put an end to the torture. The *comes et professor mechanicae*, for his part, kept maintaining that the bridge could be repaired, and that it had been Auxentius who embezzled the money.

We do not know whether or not the case was solved: our evidence from Symmachus stops here, and he turns to other matters. What emerges from this episode is first of all the obvious fact that architects could have senatorial rank and that there was such a thing as a *professor* of mechanics, a title which seems to denote some official recognition of expertise, and possibly of teaching capacities. Moreover, we are given glimpses into the several ranks at the building site: the architects first of all, trusted with administration of money (and responsibility for it); then, probably next in status, the master craftsmen, who are trusted with the important task of assessing the works when the two chief people start accusing each other; finally, the *artifices* themselves, who are again able to function in a semi-official capacity by giving a financial estimate, and the diver, someone with a degree of specialization or skill, who is interrogated simply as witness (not as expert witness) – given their low social status, such people could be tortured if their testimony sounded suspicious.

Another edifying story is told by a contemporary of Symmachus, Augustine. A pupil of his in Carthage, Alypius, had been unjustly apprehended by an angry mob with the accusation of burglary. An architect, who was the top man in charge of public buildings, happened to pass by as the student was being taken away and recognized him from having met at the house of a common acquaintance – a senator. Having listened to Alypius' version of the story, the architect "ordered all the people there, who were in an uproar and making threatening shouts, to come along with him," and led them to the house of the real offender, who was brought to justice. What I find notable in this episode is not so much that the architect was a frequent visitor to a senator's house, as that his authority with the man in the street was such that he could use it to persuade an angry crowd to do as he said.[37]

Another late witness is Cassiodorus, who, in letters dating from the sixth century, charged an architect with the task of repairing some communal baths, which involved the administration of public money, and on another occasion requested another architect for public work in Rome. In this second case, Cassiodorus reminded his addressee of the seven wonders of the ancient world and expressed the wish that the architect chosen for the job "apply himself to books" in order to stand comparison with such a glorious past.[38] We can only guess what the books in question may have been: in another letter Cassiodorus exhorts a steward of the imperial retinue to take inspiration from Euclid and Archimedes for "beautiful shapes" with which to adorn the imperial palace.[39]

Cassiodorus provides evidence for *mechanici* as well: he mentions a mechanician who had been paired with a water-

[37] *Confessiones* VI.9. Architects are mentioned as public officers in the *Codex Justiniani* 12.19.12.1 (a law of the emperor Anastasius, AD 491–518).
[38] *Variae* II.39 and VII.15 respectively.
[39] *Variae* VII.5.

diviner "so that the waves discovered by one can be lifted by the other, and what nature does not allow to go upwards can rise artificially."[40] In another letter written on behalf of King Theodoric (*ca.* 507–511), he invites two *spectabiles viri* between whom a boundary dispute has arisen to entrust their case to the capable hands of a land-surveyor, who will solve it "by means of geometrical forms and land-surveying knowledge" rather than with weapons, and will enclose the contested land as diligently as the speech is enclosed by each letter. Cassiodorus continues with a micro-history of land-surveying: how it originated with the Chaldeans, how it was taken up by the Egyptians and eventually by Augustus, who carried out an extensive programme of land-division. The "metrical author" Hero is then mentioned as the person who "made [land-surveying] into a written doctrine, so that the person involved in this study could learn by reading what he would have fully to demonstrate to the eyes." The discipline enjoys a great reputation indeed when compared to other branches of knowledge: arithmetic, Cassiodorus says, is taught to empty schoolrooms; geometry "insofar as it discusses heavenly things" is only known to scholars; astronomy and music are learnt just for their own sake, but the *agrimensor* "shows what he says, and proves what he has learnt."[41]

More light on the public persona of land-surveyors is cast by the texts collected in the so-called *Corpus Agrimensorum Romanorum*, which includes treatises ranging from the first century BC to the early fifth century AD at least, but chiefly from the second century AD.[42] Unlike most ancient scientific practices, our evidence for land-surveying is not limited to the

---

[40] *Variae* III.53, esp. 6; quoted in Oleson (1984), 34.

[41] *Variae* III.52.

[42] There are two main editions of the *Corpus* so far: one by K. Thulin includes Frontinus, Agennius Urbicus, Hyginus, Siculus Flaccus and Hyginus Gromaticus. The other, by F. Blume, K. Lachmann and A. Rudorff is complete but arguably less accurate. There is also an edition with French translation of Siculus Flaccus' *De condicionibus agrorum*, and Brian Campbell (see (1996)) is preparing an English translation of the *Corpus*. See also Gabba (1984); Hinrichs (1992); Toneatto (1992). References are to Thulin's edition unless otherwise specified.

texts themselves. We have both further testimonies in legal, historical and literary works and abundant material evidence: inscriptions recording official decisions about, say, the territorial boundaries of two neighbouring cities; boundary stones; maps; surveying instruments; tombstones of land-surveyors. Traces of land-division detectable by means of, for instance, aerial photography are numerous, especially in Italy, France, the former Yugoslavia and North Africa.[43]

Even though late antiquity hardly saw the centuriation of new territories, everyday general administration involved settling disputes about boundaries, as we have seen in the case of Cassiodorus, or division of heredity.[44] We have entries in the collections of laws known as *Codex Theodosii* and *Digesta* on the "Administration of boundaries": land-surveyors were among the main arbitrators in such cases.[45] If a dispute arose between two parties and one of the two brought in a land-surveyor, the fee had to be shared by both.[46] A *mensor* was also instrumental in cases where a flood had obliterated the borders between properties: in this case he worked for the governor of the province, i.e. in an official "public" capacity. If necessary, the governor would inspect the situation personally.[47] Another entry in another law collection, the *Codex Justiniani*, runs thus: "The chief (*primicerius*) of the land surveyors after completing two years [of service] is assigned the lowest office of *agens in rebus*."[48] And again, *mensores* and *metatores* are mentioned as part of the staff of several governmental departments – their duties included dividing prop-

---

[43] See e.g. Dilke (1971); Hinrichs (1974); Chouquer & Favory (1992).

[44] Centuriation, used loosely, means the division of a territory in squares or, less frequently, rectangles of a given size. A late example of land (re)division was with the emperor Julian, who, according to Eunapius, *Vitae sophistarum* 493, measured the land with the aim of relieving the Greeks from part of their tribute.

[45] *De finium regundorum*, in *Codex Theodosii* 2.26, with laws ranging from AD 330 to 392, and *Digesta* 10.1.

[46] *Dig.* 10.1.4.1.

[47] *Dig.* 10.1.8: "si ita res exigit, oculisque suis subiectis locis."

[48] *Cod. Just.* 12.27.1 (AD 405, addressed to the *magister officiorum*): "Primicerius mensorum biennio expleto agentis in rebus ultimi militiam sortiatur." For the use of military terms (i.e. *militia*) for civilian purposes, see Kelly (1998), 168, who also gives further bibliography on bureaucracy in the late Roman Empire.

erty for billeting purposes.[49] This picture seems to be confirmed by the late fourth-century *Notitia dignitatum*, where land-surveyors figure on the staff of both military and civilian departments.[50]

The *Digesta* also contain a heading "Si mensor falsum modum dixerit" ("If a land-surveyor declares the wrong measure") which establishes various sanctions against land-surveyors who do not do their job properly, and considers the case where instruments are used. The jurists (Ulpian and Paul) mention, as cases parallel to those of surveyors, the professional circumstances of other practitioners, namely architects and accountants (*tabularii*), against whom it was also possible to take legal action on the basis of fraud (*dolo malo*).[51] The reason why the type of legal action appropriate to the case would have been for fraud, not for a job badly done, is because the relation between a *mensor* and a customer was not a commercial one: it did not fall under the heading of *locatio conductio*, which included transactions for manual jobs or hired labour, but was defined more in terms of a favour, *opera beneficii*. Thus, the emolument of a *mensor* or of anyone who performed a favour rather than a paid job was called *honorarium* rather than *merces*, i.e. it was intended to be more a thank-you gift than remuneration. Reality was, as it happens, rather different: the entry in the *Digesta* comments that, although officially land-surveyors should not be paid, in fact

---

[49] The laws in question are *Cod. Just.* 12.40.1 (mentions "mensores nostri"); 12.40.3; 12.40.5; 12.40.9 (the latter two mention *metatores*) and 12.59.10, where they are listed as part of the staff of the *scrinium sacrorum libellorum* along with pedagogists, *cellarii* and *lampadarii*. The laws date from the fourth to the fifth century. *Metatores* seems to have been the specific term for military *mensores*: in Vegetius (*ca* AD 400), *Epitoma rei militaris* 2.7, the *metatores* go in front of the rest of the army and choose the best site for encampment, while the *mensores* and *agrimensores* divide up the camp once the site has been chosen, *ibid.* 3.8.

[50] *Notitia dignitatum Orientis* 7.66 (*mensores* in the office of the *magister militum*) and *Or.* 11.12 (with the *magister officiorum*, they share an entry with *lampadarii*).

[51] *Dig.* 11.6.7 (Ulpianus): "This action is also given against a surveyor who uses mechanical instruments, if he deceives ... By analogy, the action should also be given against an architect who deceives ... I think actions should also be given against accountants who deceive in their calculations." ("Et si mensor machinarius fefellerit, haec actio dabitur.... Hoc exemplo etiam adversus architectum actio dari debet qui fefellit ... Ego etiam adversus tabularium puto actiones dandas, qui in computatione fefellit.") On these issues see Visky (1959).

they are, but this still does not make them liable to the conditions that hold for hired labour. The legal nature of the land-surveyor's job thus seems to indicate that his task was seen, at least in principle, as quite prestigious.

Land-surveyors as represented by the authors of the *Corpus* come across as remarkably self-aware – there are several references to "professio nostra"[52] – and as knowledgeable about philosophy, geography and mathematics. Hyginus Gromaticus (second century AD) has a rare mention of Archimedes' *Arenarius*;[53] Balbus, an army-trained land-surveyor who fought with Trajan or perhaps with Domitian in the Dacian wars (or perhaps against the Germanic tribes),[54] tells his addressee Celsus that it would seem disgraceful to him if, when asked how many types of angle there are, he could only answer: "many", since it is part of their job to have more than just trivial geometrical knowledge.[55]

Marcus Junius Nipsus' treatise (second century AD, maybe later)[56] contains a number of problems which involve finding a certain element of a geometrical construction or object when some of its other elements are given. The problems are set in a specific, rather than a general form, so that the elements which are given are expressed as numbers, the procedure is by calculation of numbers, and the element which is sought will also be expressed as a specific number. For example,

[i]f a right-angled triangle is given, and the cathetus and the basis are given and are together 23 feet, and the area of this triangle is 60 feet and the hypotenuse 17 feet, [suppose it is required] to state the cathetus and the basis separately. We would find out in this way. I multiply the number of the hypotenuse by itself. It makes 289. From this I take away four areas, which makes 240. There is a rest of 49. Of this I always take the side [i.e. the

[52] E.g. Siculus Flaccus, who, according to Dilke (1971), 44, dates from the third century AD, *De condicionibus agrorum* 98.9; Balbus, *Ad Celsum expositio et ratio omnium formarum* 93.14 (Blume).

[53] Hyginus Gromaticus, *De limitibus constituendis* 148.4–7.

[54] For the uncertainties of interpretation, see Dilke (1971), 42.

[55] Balbus, *Ad Celsum* 93.11–15 (Blume): "foedum enim mihi videbatur, si genera angulorum quot sint interrogatus responderem 'multa': ideoque rerum ad professionem nostram pertinentium, in quantum potui occupatus, species qualitates condiciones modos et numeros excussi."

[56] For the dating, see Dilke (1971), 60.

square root]. It is 7. This I always add to the two together, that is, to 23. It makes 30 feet. Of this I always take the half. It is 15. This is the basis of the triangle in question. From the two together, that is, from the 23, I take away 15 feet. The rest is 8 feet. It will be the cathetus.[57]

Indeed, the mathematics of the surveyors often applies numbers to geometrical problems and makes extensive use of instruments and of "material" ways of defining concepts. For instance, Balbus describes some geometrical objects thus: "There are three kinds of lines, right, circular, curving ... A curving line is multi-shaped, like the line of ploughed fields or ridges or rivers."[58]

The land-surveyors often compare and contrast their kind of knowledge with that of other people: for instance, Agennius Urbicus (originally thought to be from the second century AD, now dated to the fourth or fifth)[59] says that the Stoics assert that the world is one, yet, if one wants to know what the world is like, and how big ("*qualis quantus*"), one needs geometrical knowledge.[60] About geometry in general, again Agennius Urbicus says:

Thus, of all the honourable arts, which are carried out naturally or proceed in imitation of nature, geometry takes the skill of reasoning as its field. It is

---

[57] Marcus Iunius Nipsus, *Podismus* 298–299 (Blume): "Si datum fuerit trigonum hortogonium, et dati fuerint cathetus et basis in se ped. XXIII, embadum huius trigoni ped. LX et hypotenusa ped. XVII dicere cathetum et basim separatim. s.q. facio hypotenusae numerum in se. fit CCLXXXVIIII. hinc tollo quattuor embada, quod fit CCXL. reliquum XLVIIII. huius semper sumo latus. fit VII. hoc semper adicio ad duas iunctas, id est ad XXIII. fiunt pedes XXX. huius semper sumo dimidiam. fit XV. erit basis eiusdem trigoni. de duabus iunctis, id est de XXIII, tollo ped. XV. reliqui ped. VIII. erit cathetus." The procedure is reminiscent of Hero's *Metrica* and, to a lesser extent, of Pappus himself (as we shall see in chapter five). Links with Hero have been suggested, and seem indeed quite likely at least for some texts in the *Corpus*, cf. Clavel-Lévêque (1992); Folkerts (1992); Guillaumin (1992); Høyrup (1996c).

[58] E.g. Balbus, 99.3–7 (Blume): "Linearum genera sunt trea, rectum, circum ferens, flexuosum ... flexuosa linea est multiformis, velut arvorum aut iugorum aut fluminum." Cf. Guillaumin (1988); Folkerts (1992).

[59] Campbell (1996), quoting an unpublished PhD dissertation by Mauro de Nardis, *The Writings of the Roman Land Surveyors: Technical and Legal Aspects*, University College London, 1994 (*non vidi* this latter).

[60] Agennius Urbicus, *De controversiis agrorum* 22.7–8. See also Santini (1990).

hard at the beginning and difficult of access, delightful in its regularity, full of beauty, unsurpassable in its effect. For with its clear processes of reasoning it illumines the field of rational thinking, so that it may be understood that geometry belongs to the arts or that the arts are from geometry.[61]

And again:

In making a judgment the land surveyor must behave like a good and just man, must not be moved by any ambition or meanness, must preserve his reputation both by his art and by his conduct.... [F]or some err because of inexperience, some because of impudence: indeed this whole business of judging requires an extraordinary man and an extraordinary practitioner.[62]

Indeed, land surveyors concur in presenting themselves not just as experts in measuring, but also as people able to resolve controversies and to bring unorderly, unmeasured space to order. Frontinus (first century AD and the object of two later commentaries) even talks about the operation of centuriating a piece of land as an expression or a restoration of its truth (*veritas*) – a truth evidently equivalent to the land becoming geometrized, as well as being brought under Roman control.[63] The presence of land surveyors in military contexts is paralleled by testimonies about other technical professions: Vegetius includes *librarii* for the keeping of accounts and *artifices* and *exercitati homines* for the responsibility for war engines in his catalogue of members of an army.[64] He also describes two methods to measure the walls of a city which an army is

---

[61] *op. cit.* 25.15–27: "Omnium igitur honestarum artium, quae sive naturaliter aguntur sive a⟨d⟩ naturae imitationem proferuntur, materiam op:inet rationis artificium geometria, principio ardua ac difficilis incessu, delectabi is ordine, plena prestantiae, effectu insuperabilis. manifestis enim rationi[bu]s executionibus declarat ⟨rat⟩ionalium materiam, ita ut geometria⟨m⟩ ine[o]sse artibus aut arte⟨s⟩ ex geometria esse intelligat⟨ur⟩."

[62] *op. cit.* 50.9–15: "in iudicando autem mensor[em] bonum virum et iustum agere debet neque ulla ambitione aut sordibus moveri, servare opinionem et arti et moribus.... quidam enim per imperitiam quidam per inpudentiam peccant: totum autem hoc iudicandi officium et hominem et artificem exigit egregium."

[63] Frontinus, *De arte mensoria* 15.7; 16.4; see Hinrichs (1992) for the most recent text (and German translation) of this passage.

[64] Vegetius, *Epitoma rei militaris* 2.7 and 4.22 respectively.

[65] *op. cit.* 4.30.

aiming to besiege: one of them involves measuring shadows. The procedure is not described in detail; Vegetius, however, claims that nobody will doubt its efficacy.[65] The leader of the army is also expected to work out how many soldiers will fit into a certain space, and how much space one should keep between rows when marching in formation without breaking ranks.[66] In fact, numeracy is one of the characteristics which should direct choice when recruiting troops. Many legions demand literate soldiers; expertise in written signs and acquaintance with reckoning and calculating are indeed to be sought, if only because soldiers take it in turns to keep records of things like days of leave awarded to their colleagues, and good numeracy helps avoid injustice and the giving of unwarranted leave with impunity.[67]

Another "technical" sector where mathematics may have played a role is public administration. The new fiscal system introduced by Diocletian, possibly around AD 297, involved periodic surveying and counting of people and resources, down to the number of cattle or of trees in a field, all in order to work out the corresponding tax liability. There is evidence that this new system was indeed enforced, at least in some parts of the empire.[68] Moreover, around AD 325–326, after the reunification of the empire under Constantine, a reform of the financial administration seems to have taken place. The prefects of the regional prefectures were made responsible for levying and distributing the *annona*, and new titles appear, such as *comites rei privatae* and *comites sacrae largitionis*. Local functionaries existed at the same time as central administrators bearing the same title.[69] An analysis of the extant evidence about the *comites rei privatae* and *comites sacrae largitionis* reveals, unsurprisingly, that the majority came from a rather high social background[70] and that several of

---

[66] *op. cit.* 3.15.
[67] *op. cit.* 2.19: "Praeter corporis robur, notarum vel computandi artem in tironibus eligendam ... notarum peritia, calculandi computandique usus."
[68] Jones (1964), I 62 f.; Bowman (1980); Williams (1985), 119 ff.; Corcoran (1996).
[69] Delmaire (1989a), 11 ff.
[70] With a couple of exceptions, see Delmaire (1989a) and (1989b), especially 97–101 and 105–111.

them had received a substantial education in rhetoric. For some there is evidence of philosophical studies.[71]

An increase in the number of people involved with the administration of finances, therefore ultimately also with things like accounts, is established. What is not clear is how important, if at all, was mathematical expertise in these contexts. The question can be considered as part of the more general matter of professionalism and bureaucracy in the late Roman Empire.[72] At least some of the public offices would seem to have required certain skills: for instance, there were departments who looked after legal cases on behalf of the Emperor, and their employees had to be trained in the law or to rely on someone who was. The departments responsible for the maintenance and construction of public buildings, including aqueducts, and the board in charge of land-surveying which was still operative in the sixth century, must have included architects and surveyors among its staff. On the face of it, it would seem that specialized skills were desirable assets that had the potential to open doors. Current thinking about late Roman bureaucracy, however, inclines to the view that access to the administration and promotion within it were to a great extent a mere matter of connections, patronage and money. Top officers tended to be educated, but that signifies more that they belonged to a certain social group than that education was indispensable or even just useful for their job.[73] In other words, according to the prevalent view, it was not necessarily the case – in fact, it was hardly ever the case – that good mathematical skills facilitated access to top positions on the financial board.

[71] For instance, Fortunatianus, *comes rei privatae* in the East from AD 370 to 377, who is defined philosopher by Libanius, *Epistulae* 694; Iovinus, who was *comes sacrae largitionis* in the East *ca* 364–365, and to whom Libanius passed on a letter by Iamblichus, see Libanius, *Epist.* 577 (AD 357); Longinianus, *comes sacrae largitionis* in the West in 399, who may have been a NeoPlatonist, see his letter to Augustine in this latter's *Epistulae* 234.

[72] Among the many contributions, see MacMullen (1964); Brunt (1975) and (1983); Bowman (1976); Pedersen (1976); Saller (1980); Kelly (1998).

[73] Pedersen (1976); Brown (1978), (1980) and (1992); Kaster (1988); Bowersock (1990).

Given that the financial board *had* to carry out some function, however, and even allowing for malfunctioning, *someone* in that department must have had some calculating skills – if not the top people, then someone employed or owned by them. The picture is thus that of a group of leading officers, non-specialized but with good connections and usually a good general education, who may have changed posts quite easily in order to move up the ladder or to another department or geographical area, plus a "hard core" of more or less skilled employees, who would stay with a department through all sorts of changes, including imperial reversals of fortune. At present, very little is known about this "hard core" of lower bureaucrats: a ground-breaking study about *notarii* and *exceptores* has showed that they were quite literate, as well as being particularly able at their specific task. There is also some evidence that some of these categories at least, especially in late antiquity, exhibited some awareness of themselves as a group – that they manifested some kind of professional identity – but further study is definitely required.[74]

The lower officers in charge of dealing with the mathematics required for administrative or commercial tasks went under several names: *calculatores, tabularii, numerarii*, λογίσται, καθολικοί.[75] Those terms are not well-defined: they could denote secretaries or clerks or administrators with no particular responsibility for account-keeping or finances; then again they may have been specifically in charge of calculations and things to do with numbers.

An early source, Martial, describes a *calculator* and a stenographer (*notarius velox*) being surrounded by many pupils, and an inscription from *ca.* AD 144 commemorates a thirteen-

[74] Teitler (1985); Kelly (1998).

[75] The examples where λογίσται are mentioned are often difficult to interpret, because the duties involved are not clear, and in small local contexts could be quite general: see *P. Oxy.* 84.2 (AD 316), a receipt from a guild of metal workers for payment by a λογιστής of a certain amount of public money for some works done for the city. Johannes Lydus (sixth century) mentions that the καθολικοί were in charge of public accounts, *De magistratibus romanis* 3.7. Inscription 657 (late third century AD) in Reynolds & Ward Perkins records a "libra[r]ius notarius [rat]iocinator n(u)m[er]arius."

year-old *calculator*, whose teacher (the dedicator of the inscription) remembers that the incredibly gifted youth wrote commentaries on the art.[76]

Tabularii many times means just notaries, or redactors of public documents who, for instance, had to be present at the moment of dividing an inheritance.[77] On the other hand, at times *tabularii* are assigned strictly financial tasks, such as calculations to do with taxes.[78] Their status must have been rather low, because they are forbidden palatine offices, and could be tortured if accused of a crime (especially in connection with abuse of their position).[79] Yet, one law exhorts a praetorian prefect to recruit *tabularii*, of whom there is a shortage, from among free men, rather than slaves or freedmen, given the importance of the office:

We prescribe by general law that, if in the whole provinces or in the individual cities *tabularii* are needed, free men be appointed and that apart from those nobody who is subject to servitude be given access to this office. And if some master has allowed his slave or tenant to deal with public acts – for we want to punish collusion, not ignorance – the person himself, to the degree to which it would benefit public utility, should pay the penalty for the computations which were dealt with by his slave or tenant, while the slave should be suitably flogged and given over to the *fiscus*.[80]

Numerarii are equated in one law with *tabularii* and share many of the latters' connotations.[81] They are also particularly

---

[76] Martial, *Epigrammata* X.62 (AD 95–8). Inscription in *ILS* 7755; see Russell (1989). For the young age of the dedicatee, see Kleijwegt (1991).

[77] *Cod. Just.* 6.22.8; 6.30.22; 7.6.1; 7.40.2; 7.72.10. On early evidence for *tabularii*, see Boulvert (1970), 421 ff.

[78] *Cod. Just.* 10.1.2 (AD 238); 10.12.49.4 (AD 382); 10.19.1 (AD 315); 10.25.1; 10.72.13.

[79] Cf. *Cod. Just.* 10.71.1 (AD 341) and 10.12.49.2 (AD 365); 10.12.49.4, respectively. There is also a communication by Valentinian III (AD 450), *Novellae* 1.3.2, where tax collectors are described as spreading terror by mixing up obscure calculations ("minutarum subputationum caligines inexplicabili obscuritate confusas").

[80] *Cod. Theod.* 8.2.5, which corresponds to *Cod. Just.* 10.71.3 (AD 401): "Generali lege sancimus, ut, sive solidis provinciis, sive singulis civitatibus necessarii fuerint tabularii, liberi homines ordinentur neque ulli deinceps ad hoc officium patescat aditus, qui sit obnoxius servituti. Sed et si quis dominorum servum suum sive colonum chartas publicas agere permiserit – consensum enim, non ignorantiam volumus obligari – ipsum quidem, in quantum interfuerit publicae utilitati, pro ratiociniis, quae servo sive colono agente tractata sunt, obnoxium adtineri, servum autem conpetentibus affectum verberibus fisco addici."

[81] *Cod. Just.* 10.12.49.2; 10.12.49.4; 10.72.13; 12.7.1; 12.28.3; 12.29.3.1; 12.49.2; 12.49.4; 12.49.6; 12.49.10; 12.49.12; 12.59.3; 12.60.6.

visible in the *Notitia dignitatum*: practically every department, either civilian or military, has a couple of them on its staff.[82]

In conclusion, it is clear that there were numerous administrative tasks that required mathematical skills, and there must have been a constant demand for people who could apply those skills in calculating and keeping balances, measuring property and constructing and maintaining public works. At present, we know very little about how their numeracy was valued or how they regarded themselves, and it is hoped that future study can cast more light on the issues I have briefly raised.

1.3 Mathematics and the law

This section will deal with official pronouncements on mathematicians and mathematical activities. We will start with Diocletian's Edict of AD 301, which was designed to fix maximum prices and wages all over the empire. It was apparently meant as a leading part of Diocletian's general economic policy,[83] its earnest intentions being underlined by the fact that it threatened with the death penalty not only hoarders and retailers, but also buyers of wares at prices higher than the ones it was meant to establish. In point of fact, the edict was probably never enforced – however, it reflects the way some professions were ranked against each other from the government's point of view, and consequently their relative status.

In the wages and salaries sector of the Edict, the highest price primary teachers could command was 50 *denarii* per pupil per month – the same as gymnastics teachers. Incidentally, 50 *denarii* was also the maximum a tailor could charge for attaching a silk neck-band – so primary teachers must have been eager to acquire a large number of pupils (or they

---

[82] See scheme at the end of Seeck's edition.
[83] Mazzarino (1951); Lauffer (1971); Williams (1985), who defines the edict as a "monumental folly", 141; Averil Cameron (1993). On salaries in late antiquity see also Alan Cameron (1965).

might just as well change profession). As for mathematics, we find *calculatores* evaluated at 75 *denarii* per pupil per month and teachers of architecture at 100, but a *geometra* shared an entry with, and was ranked the same as, a teacher of Greek and Latin grammar, at 200 *denarii*. One step higher (250 *denarii*) we find lawyers, orators and rhetoricians – the highest pay was for a lawyer, who received 250 *denarii* for opening a case and 1,000 *denarii* for pleading a case. The list aimed at completeness, but some categories, notably philosophers and doctors (but not veterinarians), are left out.[84]

Apart from the question of the extent to which the Edict was a reflection of actual reality, there is the problem of the precise meaning of some terms, especially *geometra*. It can be interpreted as someone who taught straightforward geometry, or as a land-surveyor, for which several alternative terms (*agrimensor*, *mensor*, and, by this time, *gromaticus* also) were in use.[85] In the Edict *geometra* could have either meaning: the fact that there is an entry on architecture weighs in favour of the land-surveying interpretation, yet the proximity to teachers of Greek and Latin and the pay, greater than for architecture – this latter profession carried a status possibly superior, and definitely equal, to that of land-surveyor – suggest that we may be in the presence of that rare bird, a "real" geometer. In any case, from the Edict one can infer that the status of mathematics was quite high – even its humblest category (*calculator*) fared slightly better than the lower level of educational employments, and its top-level jobs were valued highly in relation to all but the upper crust professions linked to law and rhetoric.

The next, vast body of evidence relates to the obligations mathematicians faced *qua* citizens, thus *qua* potential tax-payers. The fiscal burden in late antiquity was not principally of the monetary kind: it included civic responsibilities that ranged from sitting on the city council to supervising road

---

[84] Cf. Tod (1957).
[85] Hinrichs (1974). *Mensor* is also used to denote a "measurer" of quantities of grain (*Cod. Just.* 10.26.1, AD 364).

maintenance or water supply to being responsible for tax collections. Of course, fiscal duties were not obligatory for all. It was possible to escape certain *munera*, or even all *munera*, on various grounds, such as service in the army, public office or the nature of one's profession. Granting immunities to individuals was usually in the hands of the town administration, and, as far as privileges for "intellectuals" were concerned, every city had an assigned quota (*numerus*) of deserving people on whom it could bestow privileges, the quota depending on how large the city was. The maximum, for instance, in a metropolis like Alexandria, was fixed at ten doctors, five rhetoricians and five grammarians.[86] Rome had no fixed *numerus*, so that all teachers practising there enjoyed fiscal exemption.[87]

Immunity could be revoked – teachers and doctors could periodically be put to the test (*probatio*) in front of the town authorities in order to prove their worth, though it seems this happened only in exceptional cases.[88] For instance in AD 362 the Emperor Julian decreed that teachers needed to be authorized by an official "scrutinizing" body, but this was probably linked to his short-lived policy of excluding Christians from teaching.[89] The criteria for assessment in a *probatio* would probably have been not so much specific knowledge as other qualities that could be appreciated by the general public, for instance teaching success and popularity or a widespread reputation for expertise and capability.[90]

In AD 384 Symmachus, again in his capacity as prefect of Rome, had to supervise the process of selecting a new *archiatrus* when one of the court physicians died, and a suitable candidate presented himself. There was a law according to

---

[86] Dig. 27.1.6.

[87] Nutton (1971), 56 ff.; see *Dig.* 27.1.6.11–12; 50.5.9.

[88] Lewis (1965) has evidence that doctors, with some exception, were assessed for immunity by a committee of laymen; cf. also Nutton (1971) and (1977); Millar (1977).

[89] *Cod. Theod.* 13.3.5: "Magistros studiorum doctoresque excellere oportet moribus primum, deinde facundia.... quisque docere vult ... iudicio ordinis probatus decretum curialium mereatur optimorum conspirante consensu"; cf. *Cod. Just.* 10.53.7. Cf. also Eunapius, *Vitae*, 493 ff., for the example of the Christian rhetor Prohaeresius, on which see Penella (1990), 145.

[90] Lewis (1965); Kaster (1988).

which he had to be judged by a committee of his peers, but the reality was more complicated, as Symmachus' letter to the emperor suggests. The candidate in question wanted to have the second most important place in the hierarchical order, rather than the place the law accorded to newer members. His claim was based on his years of court service and on a special privilege from the emperor himself that gave him the place left vacant by the now dead *archiatrus*. Following both the law and usual practice ("lege et more") Symmachus summoned up the *collegium* of doctors, whose members seem to have split into two factions. The older ones did not dare to decide between the law and the personal privilege ("inter vener-ationem legis et novi beneficii reverentiam iudicare non ausi"), and found a compromise solution by offering the can-didate the place he would occupy, had he been admitted to the *collegium* the moment he entered court. The majority of members, on the other hand, defended the application of the law, "munita lege divali," citing examples of other doctors who had moved to Rome without getting any special treat-ment. In a clear case of conflict between imperial law (pro-mulgated by the emperor's revered father) and imperial au-thority, Symmachus turns to the emperor himself for advice.[91]

In general, the point of granting immunities to some cate-gories of people was that what they did was considered a contribution to the welfare of the empire, either because they educated the youth or because, as in the case of architects or shippers, they provided essential services.[92] Even though it has been argued that contemporary sources make the fiscal burden look much worse than it actually was,[93] escaping taxes seems to have been a real concern, and there are several

---

[91] The law in question is *Cod. Theod.* 13.3.9 (AD 370); Symmachus' letter at *Rela-tiones* 27. See also Millar (1983), 77; Kelly (1994) and (1998), 174.

[92] Nutton (1971) and (1977); Millar (1983); Levick (1985); Kaster (1988), 118. *Utili-tas*, whose semantic range is wider than "usefulness", was a buzz-word not only in the laws themselves, but also in late ancient treatises such as Faventinus' *De di-versis fabricis architectonicae* (third century AD); the anonymous *De rebus bellicis* (fourth century AD, on which see Baldwin (1978)) or Vegetius' *Epitoma rei militaris* (early fifth century AD). For a discussion of the issue see Ireland (1979); Wirth (1980); Fleury (1996).

[93] Hopkins (1980); Bagnall (1993), 172.

reports of citizens trying to avoid their *munera* by corruption, by feigning to belong to one of the professions which enjoyed fiscal exemption (we have cases of false shippers and bogus philosophers), by enlisting in the army or simply by running away from their cities.[94] In sum, immunity from some curial obligations and especially from "mean duties" (*sordida munera*, which included billeting soldiers) was an indication of status, and the inclusion or exclusion of privileged categories can count as an index of their degree of official recognition. For instance, from AD 313, by decree of Constantine, the Christian clergy obtained immunity.[95]

Which categories were to be regarded as worthy of distinction was a matter for continuous negotiation, and the laws abound in qualifications and additions, which deny or accord privileges to borderline groups. We have a number of regulations dating from Antoninus Pius (AD 138–161) to the prefect of Constantinople Proculus (AD 428), collected under the headings "On doctors and professors", "On professors and doctors", "On exemptions", "On duties and honours" and "On exemption and immunity from duties"[96] from which we learn, for instance, that poets could not hope for any fiscal alleviation.[97] Another endangered category was primary teachers. It is specified that they are not to be included in the privileged group of grammarians,[98] yet a decree by Vespasian and Hadrian confirms their immunity from billeting, at a point when they were already enjoying the privilege.[99]

Indeed, laws often limited themselves to ratifying (resign-

---

[94] *Cod. Just.* 12.33.2, mentioned in Jones (1964), I 69 and Millar (1983), 86. Cf. also Pricoco (1985); Bowman (1986), ch. 3.

[95] Cf. Dupont (1967); Elliott (1978); Averil Cameron (1993), 56.

[96] "De medicis et professoribus", *Cod. Theod.* 13.3; "De professoribus et medicis" *Cod. Just.* 10.52; "De excusationibus", *Dig.* 27.1.6; "De muneribus et honoribus", *Dig.* 50.4.11; "De vacatione et excusatione munerum", *Dig.* 50.5.2, 10, respectively.

[97] *Cod. Just.* 10.52.3. But cf. *P.Oxy.* 2338, a late third century papyrus which lists, for the period from AD 261–262 to 288-289, the prize-winners in certain competitions which had fiscal exemptions as one of the rewards. The list, quoted in Baldwin (1982), 79, includes poets.

[98] *Dig.* 50.4.11 and 50.5.2.8

[99] *Dig.* 50.4.18.30.

edly, one may think) what was a *de facto* situation. One entry in the *Digesta* runs thus:

> The governor of a province regularly settles the law on salaries, but only for the teachers of the liberal studies. We regard as liberal studies those which the Greeks call ἐλευθέρια. Rhetors will be included, grammarians, geometricians. The claim of doctors is the same as that of professors, perhaps even better, since they take care of men's health, professors of their pursuits. But [one must not include people] who make incantations or imprecations or ... exorcisms. For these are not branches of medicine ... But are philosophers also to be included among professors? I should not think so, not because the subject is not hallowed, but because they ought above all to claim to spurn mercenary activity.... Although also masters of an elementary school are not professors, nonetheless, the custom has arisen that cases involving them should be heard, also those involving archivists and shorthand writers and accountants or ledgerkeepers. But the governor of a province must not hear outside the regular system cases of workers or craftsmen in fields other than those involving writing or shorthand.[100]

This law is an exercise in boundary-drawing. First comes the inclusion of "liberal" studies, with a bow to the officially recognized concept of *paideia* as a Greek-speaking activity; then a clause to include doctors, but to exclude at the same time dubious healing practices. Philosophers are also excluded on moral grounds: they should not court wealth. As for primary teachers, the boundary bends, so to speak, to accommodate what has become a situation *de facto*, if not (yet) *de iure*, and takes on other borderline categories: generally literate and numerate people, necessary for administration on various levels. At the same time, it keeps out workers or artisans whose tasks have nothing to do with written signs. The

---

[100] *Dig.* 50.13.1, originally Ulpian's *De omnibus tribunalibus* 8: "Praeses provinciae de mercedibus ius dicere solet, sed praeceptoribus tantum studiorum liberalium. liberalia autem studia accipimus, quae Greci ἐλευθέρια appellant: rhetores continebuntur, grammatici, geometrae. Medicorum quoque eadem causa est quae professorum, nisi quod iustior, cum hi salutis hominum, illi studiorum curam agant ... non tamen si incantavit, si imprecatus est, si ... exorcizavit: non sunt ista medicinae genera ... An et philosophi professorum numero sint? et non putem, non quia non religiosa res est, sed quia hoc primum profiteri eos oportet mercenariam operam spernere.... Ludi quoque litterarii magistris licet non sint professores, tamen usurpatum est, ut his quoque ius dicatur: iam et librariis et notariis et calculatoribus sive tabulariis. Sed ceterarum artium opificibus sive artificibus, quae sunt extra litteras vel notas positae, nequaquam extra ordinem ius dicere praeses debebit."

criteria of inclusion seem to be prestige, utility and recognition
of an actual state of things which would be difficult to change
(which is probably also a clue to the real power of people like
librarii, notarii and *calculatores*); the criteria of exclusion reflect social and moral concerns.

Philosophers are a sorry case, legally speaking: they are the
butt of some irony in a law where the compiler reports that in
Antoninus' constitution about fiscal exemptions for teachers
of liberal arts the fixed quota for philosophers was not specified, since there were so few of them. From the point of view
of taxes, he adds, "the very wealthy will voluntarily give the
benefits of their studies to their countries, but if they direct
their arguments to their own ends, they will immediately be
revealed not to be philosophers."[101] The idea that true philosophers should have a wholesome scorn for money is deployed also by Diocletian and Maximianus in order to exhort
a recalcitrant would-be philosopher to set a good example by
the others and pay his due.[102]

On the other hand, quite a lot of evidence goes to show that
people did try to adduce the profession of philosophy as an
excuse to escape fiscal obligations. Philostratus reports two
such episodes, regarding Favorinus (AD 80–150) and Philiscus
(Caracalla's time, *ca* AD 188 to 217), and there are several
others, including at least one example from the fourth century
AD: a Silbanos who somehow managed to prove (ἀποδεῖξαι)
that he was a philosopher, in order to get fiscal immunity.[103]

[101] *Dig.* 27.1.6: οἶμαι δὲ ὅτι οἱ πλούτῳ ὑπερβάλλοντες ἐθελονταὶ παρέξουσιν τὰς ἀπὸ
τῶν χρημάτων ὠφελείας ταῖς πατρίσιν· εἰ δὲ ἀκριβολογοῖντο περὶ τὰς οὐσίας, αὐ-
τόθεν ἤδη φανεροὶ γενήσονται μὴ φιλοσοφοῦντες. The notion that philosophers
should not meddle with money is a *topos* at least as old as the Sophistic movement (and its critics). Cf. also Aelian, fr. 1038 *ap.* Eunapius, *Vitae* 461, about the
son of a tradesman who was sent to be educated in making money (ἐπὶ παιδείαν
χρηματιστικὴν) but became a philosopher instead, with the result that his father
kicked him out as useless (ἀχρεῖος).
[102] Polymnestus is told "your profession and your wish are at odds" (professio et
desiderium tuum inter se discrepant): *Dig.* 50.13.1 and *Cod. Just.* 10.42.6.
[103] Philostratus, *Vitae sophistarum* 490 and 622–623 (written *ca* third century AD);
cf. also Pliny, *Epistulae* 10.58; Dio Cassius, *Historia* 78.7.3; Aelius Aristides,
Hieros Logos 73. Silbanos in *P.Lips.* 47 (*ca* AD 372), edited in Mitteis (1906),
163–165. See discussion of many of these cases in Bowersock (1969); Griffin
(1971).

Also, a law I have already mentioned[104] exempted philosophers from billeting, and the very same Ulpianus who is the source for *Digesta* 50.13.1 (where philosophers get the thumbs down) has philosophers as well as doctors, rhetoricians and grammarians exempt from the upkeep of buildings – which remains instead an obligation for teachers of civil law and geometers.[105] My impression is that it may have been worth posing as a philosopher after all, if we have a law that thunders against those who try "to usurp the habit of philosophy improperly and insolently."[106]

Our legal evidence throws light on current notions of mathematical professions as well: at the low-status end of the spectrum we find *calculatores*, *tabularii* and *numerarii*; then here was the motley group of astronomers, astrologers and weather-persons; "technicians", such as architects, engineers and land-surveyors, and, finally, of course, teachers at different levels.

The case of astronomy shows how important it was to have a good reputation. The importance of astronomy, for instance in compiling calendars for both civic and agricultural purposes, was undisputed.[107] On the other hand, the official documents abound in scathing comments on *mathematici* – here to be taken as astrologers, who, as is known, were not sharply distinguished from "real" astronomers. We have several laws, dating from Diocletian and Maximianus (AD 294–305) to Honorius and Theodosius (AD 409), collected under the general heading "On sorcerers and astrologers and others similar."[108] In AD 319 Constantine stated that no diviner should enter someone else's house, on pain of death by being burnt alive; deportation and dispossession of all property would follow for the people who had invited him in.[109] The

---

[104] *Dig.* 50.4.18.30.
[105] Ulpianus, *De excusatione* 149.
[106] *Cod. Just.* 10.52.8, dated AD 369: "Habitum philosophiae indebite et insolenter usurpare."
[107] Cf. Clark (1971); Salzman (1990).
[108] "De maleficis et mathematicis et ceteris similibus", *Cod. Theod.* 9.16 and *Cod. Just.* 9.18.
[109] *Cod. Just.* 9.18.3.

law clearly had a loophole in that one could resort to con-
sulting the witches in their own evil abode – maybe it is for
this reason that in the course of AD 357 and 358 we have three
more laws (by Constantius and Julian) to the effect that nobody under any circumstances may consult "an haruspex or
an astrologer."[110] But some lines were drawn.

In a panegyric dated AD 362, Julian is praised for allowing
philosophy and star-gazing again.[111] Constantine had indeed
toned down his own regulation in AD 321, specifying that it
was a bad thing to use magic arts "against the well-being of
people" ("contra salutem hominum"), but that there was
nothing wrong in forecasting the weather for agricultural
purposes.[112] This suggests that even harmless star-gazers may
have been frowned upon: in fact, in Diocletian's Edict on
prices, those who tried to forecast the weather were accused of
being part of the same greedy throng whose attempts to reduce the world to misery had prompted the Edict in the first
place. They had watched the skies in the hope that it would
not rain and that the harvest would be scarce, so that they
could profit from it.[113]

All in all, it seems that the government made a distinction
between the different uses one could make of one's knowledge
– what that knowledge consisted of being somewhat immate-
rial, the difference was given by the ethical or civic qualities of
e.g. the *mathematicus*, whether he wanted to do evil or to
benefit "the divine duties and men's work."[114] Another distinction is explicitly made by Diocletian and Maximianus,
who assert "It is to the public advantage to learn and to

---

[110] *Cod. Just.* 9.18.5.6. Cf. *Cod. Just.* 1.4.10 (AD 409), where it is assumed that
mathematici are pagan recidivists, whose books should be burnt. Cf. also Mac-
Mullen (1966); Straub (1970); Grodzynski (1974).

[111] Mamertinus, *Gratiarum actio* 23.4–5.

[112] *Cod. Just.* 9.18.4.

[113] The Edict in Frank (1940), 310 ff. (esp. 312 f.): "so that without a doubt [those]
men constantly plan actually to control the very winds and weather from the
movements of the stars, and, evil as they are, they cannot endure the watering of
the fertile fields by the rains from above." This reminds one of an anecdote about
Thales, (*apud* Aristotle, *Politica* 1259a6, DK 1A10), according to which the philosopher forecast good season for olives, only to monopolize the olive-presses of
the region and hire them out at very high prices.

[114] *Cod. Just.* 9.18.4.

practice the art of geometry. But the damnable astrological art is prohibited under any circumstances."[115] While the second sentence falls in line with the other laws we have seen, the praise of geometry would suggest that this form of knowledge had a positive profile: it was seen as being publicly beneficial. But then again, what exactly is the law talking about? If the reference is to "real" geometers, it implies not only that geometry was seen as a good thing for society, but also that there was some idea of geometrical activity as something that one could not just learn (*discere*) but practice (*exercere*) – one would expect this latter to include teaching, but not to be identical with it. On the other hand, the interpretation of "ars geometriae" as "land-surveying" makes perfect sense, though it would seem strange for anyone to mix up two activities apparently so disparate as astrology and agrimensure. Perhaps the use of sighting instruments (the *diopira* described by Hero allegedly served both as a surveying and an astronomical device) and the fact that astronomical knowledge was common to both justified the association.[116] The term is less ambiguous in other contexts, for instance whenever geometry is found included among the liberal arts.[117] In those cases its profession carried an entitlement to various privileges *qua* teacher of liberal arts – exemption from billeting "among other things,"[118] and, as we have seen above, entitlement to have some cases adjudicated by the governor of the province.[119]

Now, although numerical restrictions were not necessarily strictly observed, the fact itself that there was in principle a fixed quota of teachers for whom these exemptions could apply is a signal that the privileged ones were seen as a small

---

[115] *Cod. Just.* 9.18.2 (between AD 294 and 305): "Artem geometr ae discere atque exercere publice interest. Ars autem mathematica damnabilis interdicta est omnino."

[116] Hero, *Dioptra*, 190.1–21; cf. also Panerai & Filippi (1984); Regli & Camaiora (1984).

[117] For an argument that *agrimensura* was not considered a liberal art, see Schindel (1992).

[118] According to *Dig.* 50.5.10 (collected by Paul, second century AD) and *Cod. Theod.* 13.3.1 (AD 321).

[119] *Dig.* 50.13.1.

group, an elite, access to which had to be regulated in some way. As I have already indicated, there was an active idea of a boundary between these few and all the rest, which rested on notions both of expertise and of ethical behaviour. The already cited *Dig.* 50.13.1 can be seen as a re-drawing of boundaries; then again, a regulation issued by Diocletian and Maximianus states that the exemption law includes teachers of liberal studies, but excludes calculators.[120]

The idea that a discipline could claim official recognition on the basis of its social utility probably underlies another group of laws, this time about architects and *mechanici*.[121] In a previous section I have discussed the role of land-surveyors as consultants or adjudicators in disputes; here their fiscal status is at issue. We have a heading "On the exemptions of technicians," where in three laws dated AD 334, 337 and 344 Constantine and his successors Constantius and Constans establish full immunity from *munera* for architects, doctors, painters, carpenters and thirty-three more "technical" categories, with the exhortation that, in the spare time from their activities, they teach other people, in particular their children, the profession.[122] One of these laws explicitly maintains that "one needs as many architects as possible" and exhorts the prefect of the African provinces to encourage towards that career any youths in their twenties who have had a taste of "liberales litteras."[123] The law of AD 344 describes the collective duties of "mechanicos et geometras [here to be taken as land-surveyors] et architectos" as administering boundaries, measuring (it is not specified what) and looking after aqueducts. Moreover, they are this time compelled to teach and enable others to teach in their turn: "with our pronouncement

---

[120] *Cod. Just.* 10.52.4: "Liberalium studiorum professores, non etiam calculatores continentur."

[121] I shall just note that A. H. M. Jones's explanation that "[a]rchitects, engineers and surveyors enjoyed the social standing which they were accorded because their arts were based on a theory which could only be acquired by way of a literary education" ((1964), II 1014) fails to explain why poets should not be always exempted, or why the status of philosophers is so dubious. Cf. also Westerink in Brown (1980), 30.

[122] "De excusationibus artificum", see *Cod. Theod.* 13.4 and *Cod. Just.* 10.64.

[123] *Cod. Theod.* 13.4.1 (AD 334).

we constrain them to pursue teaching and learning equally. Accordingly let them enjoy the immunities and raise enough teachers to teach in their turn."[124] The assumption is that expertise entails an ability to teach and that teaching establishes a tradition or a group, thus ensuring continuity over time; the invitation is to create such a self-perpetuating tradition, where the competence of the individual is guaranteed by his being part of the group.

But what example was set by the leaders? That is, what about the mathematical education of the top rank of political power? The *Historia Augusta*, probably composed in the 390s AD, presents several Roman emperors as possessed of extensive learning – as embodying, in short, that ideal of *paideia* which was such an important part of the self-identity of the upper classes of late antiquity. Sometimes the emperors' learning specifically included some form of mathematical knowledge. Hadrian was "very expert" (*peritissimus*) in arithmetic and geometry, and liked to surround himself with (among others) *geometrae* and astrologers.[125] Severus Alexander was expert in astrology and carried out geometry ("geometriam fecit"; both this and Hadrian's case may refer to landsurveying), as well as promoting astrologers, mechanicians and architects.[126] Septimius Severus, too, was at home in astrology.[127] The point is clearly not whether these accounts are historically accurate (they probably are not), but the fact that, to the eyes of whoever wrote the *Historia Augusta*, a certain education was part of being an emperor, and that education often included mathematics.

The philosophical leanings of the emperor Julian are well known: he studied in Athens, and echoes of Porphyry and Iamblichus have been detected in his writings. It is also well known that he upheld the idea of Hellenic *paideia* in opposi-

---

[124] *Cod. Theod.* 13.4.3: "in par studium docendi adque discendi nostro sermone perpellimus. Itaque inmunitatibus gaudeant et suscipiant docendos qui docere sufficiunt."

[125] *Scriptores Historia Augusta: Hadrianus* 16.10.

[126] *SHA. Severus Alexander* 27.5–7 and 44.4 respectively. Cf. Straus (1970).

[127] *SHA. Septimius Severus* 3.9; *Pescennius* 9.6.

tion to "new" or simply non-Hellenic traditions such as Christianity or Judaism. In his speech *Against the Galileans*, written around AD 362–363, he compares and contrasts Jewish traditions and wisdom with Hellenic ones, and has also something to say about the "men of the Western nations" (apparently, this meant Gauls and Iberians), who, unlike Romans and Hellenes, have no great inclination for philosophy or geometry or things like that.[128] When rhetorically asking when the Jewish people ever invented any science or philosophy, Julian recounts a mini-history of science which is uncannily similar not only to NeoPlatonist, but also to some twentieth-century AD reconstructions:

The theory of the heavenly bodies was perfected among the Hellenes, after the first observations had been made among the barbarians in Babylon. And the study of geometry took its rise in the measurement of the land in Egypt, and from this grew to its present importance. Arithmetic began with the Phoenician merchants, and among the Hellenes in course of time acquired the aspect of a regular science.[129]

Later on, Justinian is presented by contemporary sources as more or less directly responsible for the planning and construction of many public buildings, including the church of Hagia Sophia, where he worked in close collaboration with a group of *mechanici*.[130]

As for the state's direct involvement in educational policy, the evidence for late antiquity is often ambiguous. Opinions about whether the Roman Empire subsidized public chairs, on how wide and regular a basis, and for how long, are cor-

[128] Julian, *Contra Galilaeas* 1.131c: ἐπὶ τὸ φιλοσοφεῖν ἢ γεωμετρεῖν ἤ τι τῶν τοιούτων.
[129] *Contra Gal.* 178a–b: ἡ μὲν γὰρ περὶ τὰ φαινόμενα θεωρία παρὰ τοῖς Ἕλλησιν ἐτελειώθη, τῶν πρώτων τηρήσεων παρὰ τοῖς βαρβάροις ἐν Βαβιλῶνι γενομένων· ἡ δὲ περὶ τὴν γεωμετρίαν ἀπὸ τῆς γεωδαισίας τῆς ἐν Αἰγύπτῳ τὴν ἀρχὴν λαβοῦσα πρὸς τοσοῦτον μέγεθος ηὐξήθη· τὸ δὲ περὶ τοὺς ἀριθμοὺς ἀπὸ τῶν Φοινίκων ἐμπόρων ἀρξάμενον τέως εἰς ἐπιστήμης παρὰ τοῖς Ἕλλησι κατέστη πρόσχημα. Analogous themes are of course contained in many earlier authors, e.g. Herodotus, *Historiae* 2.109; Diodorus Siculus, *Bibliotheca* 1.69.5; 1.81.1-2, and some later ones, e.g. Proclus, *In primum Euclidis elementorum* 64.3–65.7; 136.8–12. Cf. Vitrac (1996).
[130] Procopius, *De Aedificiis* I.1.24, 71, 76; II.3.11. Cf. also Downey (1946–8) and (1947–8).

respondingly divided.[131] We do have complaints by contemporaries about the absence of support for education and claims that the employment of teachers should be a public concern, along with indications that this concern was sometimes met.[132] For instance, as Constantinople grew in stature and prestige, public teachers were appointed there.[133] Symmachus refers on various occasions to public teachers, including his own brother Priscianus, who was paid by the state to teach philosophy.[134] In a similar position was the philosopher Celsus, who, Symmachus writes, was prepared to come to Rome to teach the nobles ("erudiendis nobilibus") without any emoluments. Considering this and the fact that Celsus' reputation "nearly equalled that of Aristotle," Symmachus proposed to elevate him to the senatorial rank, with the specification that he be immune from *munia publica*.[135] The *praefectus urbis* seems to have been responsible for the educational welfare of other cities in Italy as well: when Augustine stood for a public post of teacher of rhetoric in Milan, it was to the authorities in Rome that the citizens of Milan applied for endowment of the chair.[136]

As for Alexandria, we have no conclusive evidence about the capacity in which the Museum, that symbol *par excellence* of Alexandrine culture, survived after the Ptolemies. We are told that the Emperor Claudius (AD 41–54) restored it to some magnificence.[137] Moreover, from AD 38 until the late third century, membership of the Museum of Alexandria was given as a courtesy title to some civil servants and military officials. The affiliation entailed being exempt from taxes, yet since those people, given their jobs, were probably already entitled to fiscal privileges, it seems likely that the whole point of being, at least nominally, a member of the Museum was that it

---

[131] The picture is clearer for earlier times: for instance, we know that Marcus Aurelius endowed several chairs of philosophy at Athens (AD 161–180); see Clarke (1971), with bibliography; Lynch (1972).

[132] Extensive evidence in Baldwin (1982); Bowman (1986), 160; Bagnall (1993).

[133] Müller (1910); Jones (1964), 549; Clark (1971); Averil Cameron (1993).

[134] Symmachus, *Epistulae* I.79, *ca* AD 378–380.

[135] Symmachus, *Relationes* 5, *ca* AD 384–385.

[136] Augustine, *Confessiones* 5.13.

[137] Suetonius, *Vitae Caesarum* 5.42.

constituted an honour and a recognition of high status.[138] The connected issue of whether and in what circumstances the library of Alexandria survived its golden age is likewise undecided, though we know that one or more public spaces devoted to the housing of books did exist in Alexandria in the fourth century AD and possibly later. For instance, we know that a Christian mob set the Serapeum on fire in AD 389. This building, originally a temple of Serapis, must have served as a library at least until then, because we know that the books contained in it were burnt as well. Moreover, a fourth-century rhetor, Aphthonius, mentions a building in Alexandria where books were kept and which made philosophizing available to lovers of such pursuits and stirred the entire city towards σοφία.[139] We also have evidence that, from at least the middle of the third century AD until at least the late fourth century AD support was given to the maintenance and copying of books and to the constitution of libraries by emperors and bishops in various parts of the Empire.[140]

Teaching activities in Alexandria seem to have continued more or less uninterrupted through the centuries. For many years the Egyptian metropolis had financed some public teachers independently of the central government, but we do not know with certainty until when and for what subjects.[141] For instance, opinions are divided on whether the geometer and philosopher Hypatia was an employee of the city or not, even though, whatever her job denomination, it is undeniable that she had a very high public profile.[142] Eunapius (ca AD 346–414) mentions a public school (διδασκαλεῖον κοινόν) in Alexandria whose leadership was assigned to the sophist

---

[138] Lewis (1963); Millar (1977), 97–8.
[139] Aphthonius, *Progymnasmata* 12, 40.3–7. Cf. also Canfora (1986) and (1995).
[140] *Cod. Theod.* 14.9.2 (AD 372). Cf. Cavallo (1975), 91–2, 114; (1984) and (1988); Alan Cameron (1984); Fedeli (1989).
[141] We have some evidence that there were publicly subsidized teachers, in some cases of philosophy, in Oxyrhynchus, *P.Coll.Youtie* 66 (AD 253–260); *P.Oxy.* XLII 3069 (third/fourth century AD) and in Hypselis (?), *P.Lips.* 47 (AD 372), quoted in Mitteis (1906) and Bagnall (1993). See also Müller (1910) and Fowden (1977), who thinks that the city of Pergamon funded philosophical teaching.
[142] Tannery (1880b); Marrou (1963); Toomer (1974); Alan Cameron (1990); Beretta (1994); Dzielska (1995) and especially Évrard (1977).

Magnus.[143] Recent excavations in one of the Alexandrine quarters have also revealed the existence of lecture halls, in use at least from the fourth century AD onwards.[144] About a generation after Pappus, Ammianus Marcellinus remarks that in Alexandria

the geometrical measuring-rod brings to light whatever is concealed, the stream of music is not yet wholly dried up among them, harmony is not reduced to silence, the consideration of the motion of the world and of the stars is still kept warm by some, even though they are few, and there are others who are skilled in numbers.... Moreover, medicine ... is so enriched from day to day that, although a physician's work itself indicates it, yet in place of every testimony it is enough to commend his knowledge of the art, if he has said that he was trained at Alexandria.[145]

The Egyptian city's long-lasting fame as a centre for "scientific" education is confirmed by other authors:[146] according to the philosopher Junior, or whoever is the author of the *Expositio totius mundi et gentium* (probably fourth century AD), in Alexandria one could find all sorts of philosophers and every doctrine, as well as doctors. It is notable that, of the other many places mentioned in the *Expositio*, only a few warrant a brief mention of any cultural activities: Beirut for its law schools, Tuscany because it is the cradle of the art of *haruspicium*, Crete and Sicily because their citizens are both wealthy and learned.[147] Moreover, the Egyptians are compared to the Greeks: not only is it impossible to find anyone wiser than an Egyptian, but "once, the Egyptians and the Greeks having fought about which of them should have the Museum, the Egyptians were found to be cleverer and more

---

[143] Eunapius, *Vitae* 498.

[144] Haas (1997), 17.

[145] *Res Gestae* 22.16.17–18 (*ca* AD 363): "nudatur ibi geometrico radio quicquid reconditum latet, nondumque apud eos penitus exaruit mus ca, nec harmonia conticuit, et recalet apud quosdam adhuc (licet raros) consideratio mundani motus et siderum, doctique sunt alii numeros .... Medicinae autem ... ita studia augentur in dies ut (licet opus ipsum redoleat) pro omni tamen experimento sufficiat medico ad commendandam artis auctoritatem, Alexandriae si se dixerit eruditum." (Loeb trans. with modifications.)

[146] E.g. Eunapius, Gregory of Nyssa, Libanius; see Schemmel (1909); Müller (1910); Nutton (1972), 172; Haas (1997).

[147] *Expositio totius mundi* 34; 37; 25; 56; 64; 65, respectively; cf. Fricoco (1985). The *Expositio* was originally composed in Greek, but only a Latin version survives.

accomplished and they won, and the Museum was assigned to them."[148] So much for Julian's potted history of science ...

1.4 Mathematics and the educated person

Several works have discussed education in the Greco-Roman world:[149] they differ as to their interpretation of *artes liberales* or of *quadrivium*, but they generally agree in claiming that mathematical education in late antiquity was either non-existent or subordinated to philosophy or the applied arts, such as architecture. Such mathematical teaching as there was, it is stressed, was very simple, and often almost exclusively oral; when not, it was limited to a handful of texts, namely Euclid's *Elements* (at least parts of it), Nicomachus, Ptolemy and Aratus.[150] Yet the *Collection*, itself an example of quite "advanced" mathematics, addresses on several occasions teachers or students of mathematics. How does this major piece of evidence fit, and how can it modify the picture that has been sketched above? First of all, let us rehearse some common notions about mathematical education in antiquity.

Primary education, which was more or less the same for all those who did go to school, included grammar, some Greek and Latin literary classics and some arithmetic. People who were literate were as a rule also numerate – that is, able to perform the basic arithmetical operations. Augustine had nightmarish memories of having to chant tables of addition as a schoolboy[151] and it is clear, for instance from passages in the Church Fathers, that anyone was expected to be able to count on his or her fingers.[152] We also have quite a few school

---

[148] *Expositio* 34: "[A]liquando certamine facto Aegyptiorum et Graecorum, quis eorum Musium accipiat, argutiores et perfectiores inventi Aegyptii et vicerunt, et Musium ad eos iudicatum est."

[149] See especially Marrou (1965); Clarke (1971); Kaster (1983); Hadot (1984).

[150] Cf. Clarke (1971); Pingree (1994). Of course, whether these texts can be considered "simple" is a matter of opinion.

[151] The "odiosa cantio" of *Conf.* I 13.22.

[152] On finger-numbering see Menninger (1958); Williams & Williams (1995). These latter include testimonies from Ambrose, Augustine, Irenaeus, Jerome, Pacianus and Sidonius Apollinaris.

notebooks, which may have belonged to children of trades-men or future *tabularii*, containing mathematical exercises, tables of fractions, multiplication, division or addition, or simple geometrical problems.[153]

When the pupils moved on from primary school they would begin their training in rhetoric. It is interesting that we can examine an influential rhetor's defense of the importance of mathematics: Quintilian (*ca* AD 35–100), in his textbook on the education of orators, discusses whether geometry was a useful thing to study for a future practitioner.[154] His answer is yes. First, he maintains that geometry exercises and sharpens the mind, and that, as it is commonly believed, its advantage lies in the process of learning itself, rather than in what is being learnt. Secondly, knowledge of numbers and shapes is one of the first necessities for anybody. There is therefore a sense to the notion that mathematics is as indispensable a constituent of basic education as reading and writing: for instance, it would be embarrassing for an educated man to be caught in the act of not being able to calculate on his fingers. Furthermore, for Quintilian some processes characteristic of mathematics, such as a demonstrative structure that goes from the unknown to the known, achieve that same logical cogency which is the aim of a rhetor when he tries to give persuasive force to his arguments.[155] Quintilian adduces examples where geometry is the only means to distinguish true from false belief. For instance, one may think that the time taken to circumnavigate an island is a reliable indication of how large the area of the island is. Yet, a knowledge of iso-perimetry would expose this belief as false, since two islands can have the same perimeter (the time taken to circum-navigate them would be equal) and widely different areas.

---

[153] E.g. Karpinski (1923); Boyaval (1977); Cauderlier (1978); Fowler (1987), esp. 270–279, which is a catalogue of available evidence.

[154] Quintilian, *Institutio Oratoria* I.10.34–49.

[155] Compare Proclus, *In Eucl.* 24.23 ff.: "To the theoretical arts, such as rhetoric and all those like it that function through discourse, [mathematics] contributes completeness and orderliness, by providing for them a likeness of a whole made perfect through first, intermediate, and concluding parts" (trans. here and henceforth G. R. Morrow with my modifications).

Quintilian concludes by claiming that the orator should be able to explain the causes of natural phenomena and show that nothing happens by chance; geometry, both for what it teaches about astronomy and because it explains problems otherwise impossible to understand, is the right auxiliary discipline to do just that.

The *Institutio Oratoria* contains in a nutshell a number of ideas about mathematics that had wide currency in antiquity: the notion that mathematics should be part of the educational curriculum not so much for its content as for the *habitus mentis* it engenders dates back to at least Plato and Isocrates.[156] In the *Republic*, Plato famously promoted mathematical studies as propaedeutic for the formation of his ideal rulers, the philosopher-kings.[157]

Augustine, for all his dislike of arithmetic tables, claimed that knowledge of numbers was as necessary as knowledge of words if one was barely to communicate with other human beings, because numbers impose a fixed limit on the multitude of things and allow men to grasp them.[158] Other notions associated with mathematics at a broad level seeped into non-mathematical discourse. A nearly-converted Augustine, on listening to Ambrose preaching, said that he wanted to be as certain that Ambrose's words were true as he was certain "that seven and three [were] ten."[159] Numbers figure in Augustine also when it comes to explaining the discrepancy between the number of generations from Abraham to Jesus Christ – forty in the gospel of Matthew, forty-two in that of Luke. The disagreement had been picked up for criticism by Julian, as evidence for the unreliability of the Scriptures.[160] Augustine attempts an ingenuous conciliation between the

---

[156] Cf. for instance Plato, *Leges*, 757a ff.; Isocrates, *Antidosis* 261 ff. and *Panathenaicus* 26 ff.

[157] Plato, *Res Publica* 521c–531c.

[158] Augustine, *De ordine* II 12–35.

[159] Augustine, *Conf.* VI.4.6. Why seven, three and ten should be chosen is probably to be understood on the basis of some relatively straightforward numerology: seven is the number of the days of creation; three that of Trinity; ten, the decad, could be the entire cosmos.

[160] Julian, *Contra Gal.* 253e.

two numbers: forty is rich in associations as a number often recurring in the Bible (e.g. the forty years Israel spent in the desert), but in fact, if one counts the generations recorded in Matthew in a different way, the total number is forty-two – the same as in Luke. Namely, one can count a certain "turning point" inclusively: there are in Matthew fourteen generations from Abraham to David, fourteen from David to the exile in Babylon, fourteen from Babylon to Christ if one counts Jechonias twice. He is indicated as responsible for the decline of faith in the true god on the part of Israel, i.e. for a *deflexio*, a "turning point" or angle. "What is in the angle is counted twice," claims Augustine. The final sum produces forty-two, and the reconciliation of the two gospel writers is complemented by a fine analogy between angle or turning point and Jesus as corner-stone.[161]

Calculations involving turning-points occur also in the philosopher and mathematician Iamblichus (early fourth century AD): in order to explain a mathematical proposition whose detailed content does not concern us here, he uses the image of a race-course along which numbers run, so to speak, (they increase from 1 to n) turn round (the turning-point is καμπτήρ) and come back (the numbers then decrease from n minus 1 onwards) to the goal-post.[162] Mathematical literature was known to the Christian writer Hippolytus, who sets out to refute a certain Pythagorean method of divination, according to which future events could be predicted by substituting for the letters of someone's name the corresponding numbers (Greek numerical notation used the same signs as the alphabet) and then interpreting their meaning. The numbers obtained from the substitution are called "basic numbers" (πυθμένες) and are variously divided following rules of seven or of nine. Now, Iamblichus also contains a proposition which employs πυθμένες and, more to the point, book 2 of Pappus' *Collection* is devoted, as we will see in chapter five, to

---

[161] Augustine, *De consensu evangelistarum* 2.4. Jesus as corner-stone, with reference to the Psalms, in the first letter of Peter, 2.6 ff.

[162] Iamblichus, *In Nicomachi Arithmeticam Introductionem* 75.22 ff A detailed description in Heath (1921), I 113.

multiplication exercises which entail both the substitution of πυθμένες for numbers and the substitution of numbers for the letters which form a short epigram.[163]

To stay within the Church, Origen (third century AD), when attacking Celsus, reported him as thinking that one might attain knowledge of God by means of a synthesis analogous to the one geometers talk about.[164] According to Eusebius, the bishop of Laodicea Anatolius (*ca* 303) had when in Alexandria been deemed worthy by his fellow citizens of establishing the school of the successors of Aristotle. He knew arithmetic, geometry and astronomy, as well as philosophy and rhetoric, and Eusebius also reports excerpts from Anatolius' canon on Easter, where he discussed equinoxes and ecliptic circles, and which he wrote along with some introductions to arithmetic in ten books.[165] And, on the non-Christian front, Eunapius reports how the sophist and doctor Oribasius, when sent into enemy territory as a punishment, showed a greatness of spirit not limited by place or circumstance, but stable and constant – just the way it happens with numbers and objects of mathematics.[166]

Another long-lived and wide-spread topos was that mathematics and justice were linked. Hero of Alexandria (first century AD) commented on the way divisions were carried out in human society thus:

if one wants to divide ... according to a given ratio, so that not even a grain of millet, as it were, of the proportion exceeds or falls short of the given ratio, it takes geometry alone. In geometry, in fact, there is balanced agreement, and the justice that comes from proportion; and the demonstration of these things is indisputable, something which none of the other arts or sciences guarantees.[167]

---

[163] Hippolytus, *Refutatio omnium haeresium* 4.13–14; Iamblichus, *In Nicom.* 103.10–104.13; Pappus at e.g. 2.9; 4.3. This again in Heath (1921), I 114–117.

[164] Origen, *Contra Celsum* VII.44: τῇ παρὰ τοῖς γεωμέτραις καλουμένῃ συνθέσει.

[165] Eusebius, *Historia Ecclesiastica* 7.32.6–20.

[166] Eunapius, *Vitae* 498: Ὀρειβάσιος δὲ ἐκτεθεὶς εἰς τὴν πολεμίαν, ἔδειξε τῆς ἀρετῆς τὸ μέγεθος, οὐ τόποις ὁριζομένης, οὐδὲ γραφομένης ἤθεσιν, ἀλλὰ τὸ στάσιμον καὶ μόνιμον ἐπιδεικνυμένης κατὰ τὴν ἑαυτῆς ἐνέργειαν, κἂν ἀλλαχόθι κἂν παρ' ἄλλοις φαίνηται, ὥσπερ τοὺς ἀριθμοὺς φασι καὶ τὰ μαθήματα.

[167] Hero, *Metrica* 140.3—142.2: εἰ δέ τις βούλοιτο κατὰ τὸν δοθέντα λόγον διαιρεῖν ..., ὥστε μηδὲ ὡς εἰπεῖν κέγχρον μίαν τῆς ἀναλογίας ὑπερβάλλειν ἢ ἐλλείπειν τοῦ

Much later, both Iamblichus in his commentary on Nicomachus' *Introduction to arithmetic* and the pseudo-Iamblichean *Theologia arithmetica* (a text which has been dated to the middle of the fourth century AD), in the course of discussing the number five, present a complex analogy between a balance suspended by its middle point, the first nine integers arranged along either arm, and justice.

The number five is seen as the mid-point of the row of numbers and as the fulcrum of the balance. The numbers on one side, i.e. from six to nine, are "heavier" than the numbers on the other side, i.e. one to four, so that the balance will not be in equilibrium. When the balance beam is suspended, the parts which exceed (i.e. the arm with numbers six to nine) will also form an obtuse, or "excessive", angle with the line of the horizon, whereas the parts which are defective will form an acute, or "defective", angle. The right angle, on the other hand, "has the principle of maximum equality." Analogously, in a case of injustice those who do the wrong are seen as being on the side of the obtuse angle, and sinking down they get progressively away from the right, which is in the middle. The wronged party, on the other hand, increasingly approach and come near the mean as they go up. In sum, the bad sink and the victims go up to the divinity.[168]

A fourth-century scholion to a passage in Aelius Aristides' *Pro Quattuor*, where the author says that Pericles honoured proportion and that geometry and proportion are beautiful, adds that the motto "Let nobody ungeometrical enter" was carved on the gates of Plato's school. The scholion also explains that ungeometrical is to be taken as "unfair or unjust. Geometry, in fact, preserves equality and justice."[169] This is a testimony to the popularity of the tradition according to which Plato's Academy forbade geometrically uneducated

δοθέντος λόγου, μόνης προσδεήσεται γεωμετρίας· ἐν ᾗ ἐφαρμογῇ μὲν ἴση, τῇ δὲ ἀναλογίᾳ δικαιοσύνη, ἡ δὲ περὶ τούτων ἀπόδειξις ἀναμφισβήτητος, ὅπερ τῶν ἄλλων τεχνῶν ἢ ἐπιστημῶν οὐδεμία ὑπισχνεῖται. The same issues in Plato, *Leges* 757a ff. and cf. Harvey (1965).

[168] Iamblichus, *In Nicom.* 17.6 ff.; [Iamblichus], *Theologia Arithmetica* 37.10 ff.

[169] Aelius Aristides, *Pro Quattuor*. *Scholia* 125.14: ἄνισος καὶ ἄδικος. ἡ γὰρ γεωμετρία τὴν ἰσότητα καὶ τὴν δικαιοσύνην τηρεῖ. Cf. also Saffrey (1968).

people to go through its gates. Although the story is very likely legendary, mathematics was considered quite generally to be closely associated with philosophy.[170] Lactantius (first decade of fourth century AD), in discussing the opportunity of education for women, slaves and manual labourers, does not exclude them from learning geometry, music and astronomy, because these subjects have some alliance with philosophy.[171]

We know of people reputed to be "real" mathematicians who were active in philosophy too: principally (and famously) Hypatia, but then again a philosopher, Hierius, to whom a (rather complex) solution to the problem of the two mean proportionals had been sent for an opinion;[172] Serenus of Antinopolis (third to fourth century AD), who is defined as a Platonic philosopher in one manuscript,[173] and Pappus himself, who shows a considerable knowledge of Plato's works in his commentary on Euclid's book 10 of the *Elements*.[174] Eutocius of Ascalona (sixth century AD), author of very well informed commentaries on Archimedes and Apollonius, also wrote (and probably taught) on Aristotelian philosophy, possibly in Alexandria.[175]

Attendance at a philosophical school, especially in late antiquity, implied a strong sense of belonging to a certain group or a certain tradition. We find an interest in constructing such groups and traditions, first of all by emphasizing the role of the teacher, the fact that he was not only an expert, but had a particular life-style that made him exemplary from an ethical point of view (he was a hero);[176] and secondly by establishing connections between past and present. Earlier prestigious au-

---

[170] Evidence collected in Hadot (1984); see also Burkert (1962); O'Meara (1989); Russell (1989).

[171] Lactantius, *Divinae Institutiones* 3.25.

[172] In book 3 of the *Collection*, 30.25–32.2.

[173] See Cauderlier (1978), 55.

[174] He mentions the *Theaetetus*, the *Laws* and the *Parmenides* (63, 64 and 76, respectively). Plato's *Parmenides* was considered part of the advanced course of philosophical studies, e.g. Dillon in Brown (1980), 23. The *Suda* also refers to Pappus as "the philosopher", *ad vocem*.

[175] Cf. Tannery (1884a); Westerink (1961).

[176] To use terms employed by Brown (1980). See also e.g. Fowden (1982); Cox (1983).

thors, especially Plato or Pythagoras, were claimed as intellectual ancestors; biographies of famous teachers were written (apart from several lives of Pythagoras, we have lives of Plotinus and Proclus); genealogies were established and continuities emphasized.[177]

Mathematics played more than one role in the philosophies of late antiquity. As a form of knowledge, mathematics was a fundamental part of their ontologies: numerical principles such as the monad and the dyad are used to explain the relation between the different parts of the universe, or the way the κόσμος came to be; mathematical demonstration serves as model for correct philosophical reasoning; mathematical disciplines are viewed as the steps of a ladder ascending to perfect knowledge.[178] Authors such as Iamblichus, Proclus, and his pupil Marinus (fifth century AD) wrote mathematical commentaries, on Nicomachus' *Introduction to arithmetic*, Euclid's *Elements* and Euclid's *Data*, respectively.

These issues have been much studied, and details need not be repeated here.[179] I would just like to sketch some aspects of mathematics as seen by some late ancient philosophers.

Boundaries are drawn between right and wrong notions of what mathematics is all about. Proclus replies to those who think that mathematics should be valued on the basis of its utility ("Land-surveying, they say, is more useful than geometry, the arithmetic of the many more than what is used as foundation, and navigational astronomy more than what is proved in general") with a discussion of just what utility mathematics should be measured against. He reaches the conclusion that mathematical knowledge should not be sub-

---

[177] On the formation of traditions cf. e.g. Manuli (1983); Dillon (1982); Donini (1982); Cox (1983); Armstrong (1984); Cambiano (1985a), (1988a) and (1988b); Cracco Ruggini (1985) and (1993); Forlin Patrucco (1985); Giuffrida (1985); Romano (1985) and (1994); Cavallo (1988); Kaster (1988); Pecere (1990); Penella (1990); Salzman (1990); Gara (1992); Averil Cameron (1993); Lloyd (1993); Barton (1994b); Lim (1995); Zanker (1995).

[178] See e.g. Plotinus, *Enneads* 1.3.3; Iamblichus, *Adhortatio ad philosophiam* symbol 36; Proclus, *In Eucl.* 21.19 ff.

[179] Among the most recent contributions see Levin (1975); Kordig (1982); Mueller (1987a) and (1987b); Napolitano Valditara (1988); O'Meara (1989); Schmitz (1997).

ordinate to human needs, but only to the needs of intellectual insight. One need not spell out the social overtones or the Aristotelian echoes of statements such as: "in general it was when men had ceased to be anxious about the necessities of life that they turned to the study of mathematics."[180] The very same Proclus, however, declares later on in the same text:

[Geometry] has devised instruments of war and defenses for our cities, made familiar the succession of the seasons and the lie of various regions, taught how to measure distances by land or sea, constructed balances and scales for determining arithmetical equality when a city needs it, invented models for exhibiting the order of the whole heaven, and many things incredible to men it has unveiled and made credible to all.[181]

Mathematics is indeed seen as having a public role, and associated with ethical qualities: we have already mentioned the simile of the balance described by Iamblichus and picked up by the anonymous fourth-century author of the *Theologia arithmetica*. Proclus echoes themes that we have seen in Quintilian and Hero when he says:

For it has been proved that areas can be unequal when [the perimeters] are equal and equal when they are unequal. A misconception is held by geographers who infer the size of a city from the length of its walls. And the participants in a division of land have sometimes misled their partners in the distribution by misusing the longer boundary line; having acquired a lot with a longer periphery, they have later exchanged it for lands with a shorter boundary, and so they have gained a reputation for superior honesty while getting more than their fellow colonists.

And again, Proclus, quoting Platonic writings, claims that mathematics "perfects us for moral philosophy by instilling order and harmonious living into our characters ... and shows up by contrast the excesses and deficiencies of vice."[182]

Mathematics is constructed as a tradition: Iamblichus often refers to the treatment the ancients gave of a certain topic and compares and contrasts it with what the mathematicians after

---

[180] Proclus, *In Eucl.* 25.18–23 (οἱ δὲ χρησιμωτέρας τὰς τῶν αἰσθητῶν ἐμπειρίας ἀποφαίνονται τῶν ἐν αὐτῇ καθόλου θεωρουμένων, οἷον γεωδεσίαν γεωμετρίας, καὶ τὴν τῶν πολλῶν ἀριθμητικὴν τῆς ἐν θεωρήμασιν ὑφεστώσης, καὶ τὴν ναυτικὴν ἀστρολογίαν τῆς καθόλου δεικνυούσης) and 29.1–3, respectively.
[181] *op. cit.* 63.9–18.
[182] Proclus, *In Eucl.* 403.4–14 and 24.4–13, respectively.

the ancients did, or (at a still further remove) with "the most recent [mathematicians]." For instance, in a discussion centred mainly on different types of ratio, Iamblichus abounds in references to Pythagoras, Plato (often claiming that things Plato said had been said before him by Pythagoras or the Pythagoreans),[183] and people with Pythagorean affiliations, true or alleged, such as Archytas, Ippasus of Metapontum, Philolaus, Eudoxus "the Pythagorean." When Pappus deals with the very same topic, *his* sources turn out to be Nicomachus "the Pythagorean" and Plato, and Ptolemy and Eratosthenes as well.[184] Even more poignantly, in the course of his account Iamblichus criticizes Euclid, whose erroneous treatment of some issues compares unfavourably, in his view, with that of some unidentified Pythagoreans.[185] As we shall see later, although Pappus makes no bones about criticizing some of his predecessors, Euclid is for him nearly a father-figure that deserves the greatest respect. Thus, evidently the past described (constructed) by Iamblichus is not the same as the past constructed by Pappus, although in at least some cases they are relying on the same sources for their information

Conclusion

To recapitulate, then, our evidence reveals that, in the case of several activities, some connection with mathematics or geometry was quintessential to the public identity of their practitioners.

Not just mathematical activities, but "intellectual" activities as a whole emerge as groups delimited by boundaries, although these boundaries may be blurred and in need of being constantly redrawn. Dividing lines do not seem to be determined exclusively by the type or extent of knowledge that a person possesses; they also depend on what use such

---

[183] E.g. Iamblichus, *In Nicom.* 105.2–11.
[184] In book 3 of the *Collection*, 68.17 ff.
[185] Iamblichus, *In Nicom.* 20.10–14; 23.18 ff. (προδηλότερον ἁμάρτημα); 24.14–17; 25.24; 30.28; 74.23 ff. Cf. also Tannery (1884b).

knowledge is put to, on the *ethos* that accompanies the profession (we have seen how astrologers and land-surveyors were busy creating one for themselves), and on its *utilitas* (interpreted in a wide sense as advantage, common benefit).

The general impression is that, far from being invisible or confined to ivory towers, mathematics did have a public profile. It figured rather prominently in official laws, it cropped up in everyday language, some jobs linked to it were characterized by the actors themselves as having a substantial degree of social visibility. This may also imply that the mathematical domain was not unaffected by some general cultural trends which we have observed in contemporary forms of knowledge: for instance, an emphasis on social and moral qualities as part of one's professionality; an emphasis on belonging to a group or tradition; a consequent interest in the past of one's discipline and an interest in definitions, classifications, and rearrangements of previous knowledge; a predilection for commentaries and epitomes as literary genres.

The next chapters will examine the *Collection* more closely and will show that it confirms this original picture in more than one respect.

CHAPTER 2

BEES AND PHILOSOPHERS

Book 5 of the *Collection* will be our first specimen. It opens with a philosophizing introduction:

Though god has given to human beings, my dear Megethion, the best and most complete understanding of wisdom and mathematics, he has allotted a part to some of the unreasoning animals as well. To human beings, as they have reason, he granted that they should do everything with reason and demonstration, but to the other reasonless animals he gave only this gift, that each of them, in accordance with a certain natural forethought, should have what is necessary for life. This gift may be observed to exist in many other species of animals, but it is specially marked among bees. Their good order and their obedience to the queens who rule in their communities are truly wondrous, but much more wondrous still is their love of honour, their cleanliness in the gathering of honey, and the forethought and domestic care they give to its protection. Entrusted, no doubt, with the task of bringing from the gods to well-educated human beings a share of ambrosia in this form, they do not think it proper to pour it carelessly into earth or wood or any other shapeless and orderless material, but, collecting the fairest parts of the sweetest flowers growing on the earth, from them they build for the reception of the honey the vessels called honeycombs, all equal to each other, similar and adjacent, and hexagonal in shape. This is how we can tell that they have contrived this in accordance with a certain geometrical forethought; the thought that the shapes must fit with each other and have their sides in common so that other things may not enter the interstices and damage their work.... Bees, then, know only what is useful for them, that the hexagon is greater than the square and the triangle and will hold more honey for the same expenditure of material in constructing each. But we, professing to have a greater share in wisdom than the bees, will aim at something rather more elaborate.[1]

---

[1] 304.1–308.8: Σοφίας καὶ μαθημάτων ἔννοιαν ἀρίστην μὲν καὶ τελειοτάτην ἀνθρώποις θεὸς ἔδωκεν, ὦ κράτιστε Μεγέθιον, ἐκ μέρους δέ που καὶ τῶν ἀλόγων ζῴων μοῖραν ἀπένειμέν τισιν. ἀνθρώποις μὲν οὖν ἅτε λογικοῖς οὖσι τὸ μετὰ λόγου καὶ ἀποδείξεως παρέσχεν ἕκαστα ποιεῖν, τοῖς δὲ λοιποῖς ζῴοις ἄνευ λόγου τὸ χρήσιμον καὶ βιωφελὲς αὐτὸ μόνον κατά τινα φυσικὴν πρόνοιαν ἑκάστοις ἔχειν ἐδωρήσατο τοῦτο δὲ μάθοι τις ἂν ὑπάρχον καὶ ἐν ἑτέροις μὲν πλείστοις γένεσιν τῶν ζῴων, οὐχ ἥκιστα δὲ κἂν ταῖς μελίσσαις· ἥ τε γὰρ εὐταξία καὶ πρὸς τὰς ἡγουμένας τῆς ἐν αὐταῖς πολιτείας

The rather more elaborate something in question is the problem of isoperimetry, or of how the contour, surface and volume of a geometrical figure relate to each other. A topic traditionally linked to this is that of the five regular solids, also known as the five Platonic bodies (pyramid, cube, octahedron, dodecahedron and icosahedron), i.e. polyhedra whose faces are equal regular polygons. Isoperimetry and the five Platonic bodies were topics known not just to geometers, but also to philosophers and, to some extent, to the wider educated public, as we have seen in the case of Quintilian and will see in greater detail in the course of the present chapter.

I will argue that one of the main reasons why Pappus chose these two topics for treatment was precisely that they were known to the wider educated public. In my opinion, book 5 addressed quite a general audience, in order not only to inform them about certain mathematical issues, but also to promote mathematics as a form of knowledge and mathematicians as the proper representatives of that form of knowledge. That is to say, and book 5 says it quite forcefully, mathematicians are better qualified than philosophers (for example) to talk about isoperimetry and the five Platonic bodies. Thus, Pappus' account is also a way of drawing boundaries between what he is doing with regard to some issues, and what other people are doing with regard to the same

εὐπείθεια θαυμαστή τις, ἥ τε φιλοτιμία καὶ καθαριότης ἡ περὶ τὴν τοῦ μέλιτος συναγωγὴν καὶ ἡ περὶ τὴν φυλακὴν αὐτοῦ πρόνοια καὶ οἰκονομία πολὺ μᾶλλον θαυμασιωτέρα. πεπιστευμέναι γὰρ, ὡς εἰκός, παρὰ θεῶν κομίζειν τοῖς τῶν ἀνθρώπων μουσικοῖς τῆς ἀμβροσίας ἀπόμοιράν τινα ταύτην οὐ μάτην ἐκχεῖν εἰς γῆν καὶ ξύλον ἤ τινα ἑτέραν ἀσχήμονα καὶ ἄτακτον ὕλην ἠξίωσαν, ἀλλ᾽ ἐκ τῶν ἡδίστων ἐπὶ γῆς φυομένων ἀνθέων συνάγουσαι τὰ κάλλιστα κατασκευάζουσιν ἐκ τούτων ἐς τὴν τοῦ μέλιτος ὑποδοχὴν ἀγγεῖα τὰ καλούμενα κηρία πάντα μὲν ἀλλήλοις ἴσα καὶ ὅμοια καὶ παρακείμενα, τῷ δὲ σχήματι ἑξάγωνα. τοῦτο δ᾽ ὅτι κατά τινα γεωμετρικὴν μηχανῶνται πρόνοιαν οὕτως ἂν μάθοιμεν. πάντως μὲν γὰρ ᾤοντο δεῖν τὰ σχήματα παρακεῖσθαί τε ἀλλήλοις καὶ κοινωνεῖν κατὰ τὰς πλευράς, ἵνα μὴ τοῖς μεταξὺ παραπληρώμασιν ἐμπίπτοντά τινα ἕτερα λυμήνηται αὐτῶν τὰ ἔργα.... Καὶ αἱ μέλισσαι μὲν τὸ χρήσιμον αὐταῖς ἐπίστανται μόνον τοῦθ᾽ ὅτι τὸ ἑξάγωνον τοῦ τετραγώνου καὶ τοῦ τριγώνου μεῖζόν ἐστιν καὶ χωρῆσαι δύναται πλεῖον μέλι τῆς ἴσης εἰς τὴν ἑκάστου κατασκευὴν ἀναλισκομένης ὕλης, ἡμεῖς δὲ πλέον τῶν μελισσῶν σοφίας μέρος ἔχειν ὑπισχνούμενοι ζητήσομέν τι καὶ περισσότερον. (Trans. Bulmer-Thomas (1967) with my modifications.) Notice the contrast between the knowledge of the bees, which is of the χρήσιμον, and that of human beings, who can afford to investigate the not strictly necessary (περιττόν).

issues. I think my interpretation is supported not only by Pappus' asides, but also by the structure itself of the mathematical argumentation.

2.1 Comparisons

Pappus occasionally refers to what he is doing in book 5 as a "comparison" (e.g. a theorem contained in the account is συγκριτικόν).[2] The term had quite general usage in ancient mathematical texts: it is found in the *Elements* in relation to the sides of the five regular solids, and it denoted in Archimedes the procedure by means of which a figure of unknown area was compared with, and eventually found equal to, a figure whose area was already known.[3]

The comparisons in book 5 are between figures with equal contour or surface, in order to establish which one has the greatest area or volume. We will in our turn compare the account in the *Collection* with analogous reports by earlier mathematicians. Apart from Plato's *Timaeus*, to which the five bodies owe their name, but whose mathematical content is too vague for our purposes, we have discussions of those themes in Euclid's *Elements*, book thirteen,[4] in Archimedes' *Sphere and Cylinder*, in Zenodorus' *On isoperimetric figures* (probably early second century BC),[5] in Hypsicles' *Book XIV*

---

[2] 348.2; see also e.g. 362.17; 410.24.
[3] Euclid, *El.* XIII.18; Archimedes, *Methodus* 428.11, respectively Cf. Mugler (1958) *ad vocem*.
[4] An anonymous writer, *In Euclidem Scholia* XIII 291.1–9, says, endorsed by Sachs (1917): "In this book, that is in the thirteenth book, he [Euclid] demonstrates the so-called five bodies of Plato, which however are not of Plato. Three of the above-said five bodies are of the Pythagoreans, that is the cube, the pyramid and the dodecahedron, while the octahedron and the icosahedron are of Theaetetus." ('Εν τούτω τῷ βιβλίω, τουτέστι τῷ ιγ', γράφεται τὰ λεγόμενα Πλάτωνος ε̄ σχήματα, ἃ αὐτοῦ μὲν οὐκ ἔστιν, τρία δὲ τῶν προειρημένων ε̄ σχημάτων τῶν Πυθαγορείων ἐστίν, ὅ τε κύβος καὶ ἡ πυραμὶς καὶ τὸ δωδεκάεδρον, Θεαιτήτου δὲ τό ε̄ ὀκτάεδρον καὶ τὸ εἰκοσάεδρον.) The entry on Theaetetus in the *Suda* lexicon says that he was the first to "construct" or "write about" the five solids (the term ἔγραψε could mean both); see the discussion in Heath (1921), I 159.
[5] Schmidt (1901) and Toomer (1972) think that he lived in the second century BC; Sarton (1938) places him between 200 BC and AD 90; Mogenet (1961) goes for a third-century AD dating.

of the Elements (second half of the second century BC)[6] and, after Pappus' time, in an anonymous appendix to Ptolemy which has been attributed to Eutocius.[7]

Although there is a good degree of overlap and cross-referring, the existence of several accounts does not mean that the issues were "standardized." For instance, the terminology fluctuated: rather casual expressions for "cube" (κύβος) and "pyramid" (πυραμίς) were at times used instead of the more precise terms "hexahedron" and "tetrahedron" – a fact underlined by Hero.[8] Also, Archimedes, according to Hero and Pappus, discovered thirteen semi-regular solids, i.e. solids whose faces are regular polygons, but not all equal. Thus, for instance, there is a type of octahedron whose faces are four regular hexagons and four regular triangles.[9] Archimedes' solids could be seen as stemming from a reflection on Euclid's account. Indeed, the statement in *Elements* XIII 18 that there are five and only five regular solids rests on a certain definition of regular solid, and is contradicted by, for instance, a solid formed by two regular pyramids glued at the base.[10] In the same line, the author of the anonymous appendix to Ptolemy remarked that one still lacked a proof that the sphere was the greatest of all solid bodies with the same surface, *including* the bodies not inscribable in the sphere. "It is left, since it is necessary to prove it, [he says] that the sphere is greater than the bodies which cannot be inscribed in a sphere as well, and our philosopher does not add it."[11]

For his account in book 5, Pappus draws on Euclid and Archimedes explicitly – indeed, he often refers to them by title

---

[6] Hypsicles, *Elementa XIV* 2–4 mentions a work by Apollonius "about the comparison of the dodecahedron and the icosahedron" which circulated in two versions, and a work by Aristaeus on the "comparison of the five bodies", *ibid.* 6.21–25.

[7] See Mogenet (1956); Busard (1980).

[8] Hero, *Definitiones* 62.24 f., 64.2 f.; cf. Waterhouse (1972).

[9] Hero, *Def.* 64.19–66.13; Pappus at 352.10 ff.

[10] Mueller (1981), 302.

[11] In the *Collection*, 1164.15 ff.: Λοιπὸν δὲ ἀναγκαίου ὄντος τοῦ δειχθῆναι [αὐτὴν] καὶ τῶν μὴ σφαίρᾳ περιλαμβανομένων μείζονα τὴν σφαῖραν, οὐδὲν προσέθηκεν ὁ ἡμέτερος φιλόσοφος. It is not clear who the philosopher in question is: perhaps Theon, perhaps Pappus.

of the book or number of the chapter, and sometimes even by number of the proposition. He also often cites Theodosius of Cyrene, a contemporary of Hypsicles and author of a *Sphaerics*.[12] On the other hand, Pappus never mentions Zenodorus or Hypsicles. If, however, we compare some of his results with some of theirs, the similarities are striking. How are we to explain Pappus' silence? It is possible (but highly improbable) that the similarities are a mere coincidence. It could also be that he knew their results through an anthology of some kind, but not their names. This second hypothesis is more likely than the first one, but there are some problems with it as well. From what we can see in the *Collection*, Pappus had access to a good number of little-known sources; indeed, we know of many otherwise totally unknown mathematicians only through him. It would be a bit of a coincidence if, of all the little-known sources, these particular two had come down to him anonymously. Moreover, Theon, who, if not directly linked to Pappus by a teacher-pupil relationship, seems at least to have had access to similar textual resources, quotes Zenodorus – is, in fact, our principal source for Zenodorus' treatise. The likeliest explanation is then that Pappus *decided* not to mention Zenodorus and Hypsicles explicitly. I assume that he must have had his reasons, if only because on other occasions, as we have said, he does cite obscure mathematicians. But more of Pappus' reasons later.

We can summarize the mathematical contents of book 5 thus:

> Among plane figures with the same perimeter, the circle has the greatest area.
>
> Among solid figures with the same area, the sphere has the greatest volume.
>
> There are five and only five regular solids.

---

[12] Theodosius is extensively mentioned in book 6, e.g. 488.26–518.15; 546.33; 616.9. Sachs (1917), 108–9 and notes, is of the opinion that Pappus' treatment of the five Platonic bodies, with its characteristic appeals to Theodosius' *Sphaerics*, was based on Aristaeus (the same person mentioned in book 7, 676.25). I do not see the necessity for Sachs's attribution, especially as it is possible to give a perfectly plausible interpretation of the text without it.

Apart from these three main results, there are several secondary results, which Pappus proves in connection with strings of auxiliary lemmas. Sometimes lemmas necessary for a proof come after, rather than before, the proposition they are supporting. That is, the proof of the proposition assumes that they are valid, and they are subsequently proved to be so. Pappus starts off stating the first main result:

1 Given any two regular plane polygons with the same perimeter, the one with the greater number of angles has the greater area,[13] and consequently
2 Given a regular plane polygon and a circle with the same perimeter, the circle has the greater area.[14]

Both statements are contained in Zenodorus.[15] Pappus' proof of 1 is along the same lines as Zenodorus', but avoids a few steps by appealing to Theodosius' *Sphaerics*. The proof of 2 is again basically the same as Zenodorus': they both quote Archimedes, from *Sphere and Cylinder* and from *Measurement of the Circle*. That is, the proof of 2 assumes that

3 A circle has the same area as a right-angled triangle whose base is equal to the radius and whose height is equal to the circumference of the circle.[16]

This formulation is found both in Archimedes and in Zenodorus.[17] In its explicit appeal to "the method in the twelfth of the elements,"[18] Pappus' proof is similar to Zenodorus', where the same step is found. Archimedes was usually recognized as the first to have discovered the area of the circle, and Pappus acknowledges his authorship of the theorem, but then disclaims direct use of Archimedes as a source. Comparison of the proof in the *Collection* with Archimedes' proof bears out that Pappus has not directly drawn on him, even though our extant text of *Measurement of the Circle*, a book which has

---

[13] 308.9–310.23.
[14] 310.24–312.24.
[15] Zenodorus, *De isoperimetris figuris* 356.1–357.22 and 358.12–360.3, respectively.
[16] 314.4–316.17.
[17] Archimedes, *Measurement of the Circle* 1 and Zenodorus, *De isop. fig.* 362.11–364.11, respectively.
[18] 314.8–15. The result corresponds to Euclid, *El.* XII.2 of our text and is about the possibility of inscribing a certain polygon in a circle.

undergone much later rewriting,[19] hardly allows serious textual comparisons.

At this point, the first main result has been proved, and Pappus proceeds with the first "excursus." He wants to prove that

4 Of isoperimetric polygons with the same number cf sides, a regular polygon is greater than an irregular one.[20]

In the case of triangles, for instance, this means that an equilateral triangle will be greater than an isosceles one, and this latter will be greater than a scalene one with the same perimeter. Pappus states this theorem, but only proves it in full after he has provided six auxiliary lemmas, which are already in Zenodorus, but with some differences. First of all, the sequence in which they are reproduced in Pappus is different. For instance, the proof of the fourth lemma in the *Collection* relies on two propositions which come *after* it in the argumentative order.[21] That implies that Pappus has to affirm something in the course of the demonstration without justifying it until later.

Again, in order to prove that, given a triangle, it is possible to construct another triangle with the same perimeter as the first one and that, on the condition of isoperimetry, an isosceles triangle will be greater than a scalene one, Pappus has two separate propositions where Zenodorus only had one. This because the first proposition, which will be used to support the second proposition, makes explicit the conditions under which a certain construction will be possible.[22]

The fifth lemma is as follows: having constructed two isosceles triangles whose bases are unequal but whose other sides are all equal to each other, it is required to construct on the same bases two other isosceles and similar triangles, so that the sum of their equal sides is equal to the sum of the equal sides of the first two triangles. This problem is found in Zen-

---

[19] Knorr (1986b) and (1989), part III.

[20] 316.18–24, corresponding to Zenodorus, *De isop. fig.* 372.4–374.7 (first stated at 364.9–11).

[21] 322.21–328.6.

[22] 316.26–322.6, corresponding to Zenodorus, *De isop. fig.* 364.12–366.2.

odorus, but again there is a difference. Zenodorus only considers the case where the first triangle is greater than the second one. Pappus, on the other hand, examines three subcases, in a passage which Hultsch considered spurious, without giving any reason for his opinion. Yet, as we shall emphasize in chapter five of the present volume, attention to specific cases is characteristic of Pappus' practice.[23]

The proof of 4, which is given at the end of the string of auxiliary lemmas, is along the same lines as Zenodorus', from which it differs in that it specifically refers to the lemmas themselves.[24] Pappus also takes this opportunity to repeat the first main result (i.e. 2): the circle is basically seen as a sort of limit-case polygon, and in particular as the most regular polygon of all, since it is as equilateral and equiangular as possible. At this point Zenodorus goes on to prove that the sphere is the greatest of all solids with the same surface, i.e. what in book 5 is the second main result.[25] Pappus, for his part, starts another "excursus", with the aim of proving that

5 Given any segments of circle with the same circumference, the semicircle has the greatest area.[26]

The theorem is a sort of two-dimensional counterpart to Archimedes' result that the hemisphere is the greatest of all sphaeric sectors with the same area;[27] its proof is again postponed to a string of auxiliary lemmas. Again Pappus mentions Theodosius in order to skip passages, and distinctions are again drawn out between specific cases. For instance, in the second lemma, about ratios of arcs to circumferences and of sectors to circles, Pappus separates commensurable and incommensurable cases.[28] The proof of 5 itself distinguishes between two subcases, in a pattern similar to Archimedes' formula for the area of a spherical segment, except for the fact

[23] 328.7–332.11 (the alleged interpolation at 330.6–332.10). The correspondent passage in Zenodorus is at De isop. fig. 366.3–367.12.
[24] 332.12–334.17.
[25] Zenodorus, De isop. fig. 374.12 ff.
[26] 334.23–24.
[27] Archimedes, Sph. Cyl. II 9.
[28] 336.26–340.10.

that here the two cases are explicitly seen as parts of a single problem.[29] The auxiliary lemmas are again referred to by means of their specific position in the text (e.g. "because of the lemma before the last but one").[30]

Pappus then moves on to the second main result, but not before having described in great detail Archimedes' thirteen semi-regular polyhedra. He also provides a method to calculate, given the number and species of the polygons that compose them, how many angles and how many sides each of the thirteen polyhedra will have, and applies it to all thirteen.[31] Having stated that

6 There are only five regular solid bodies[32] (which is the third main result and will not be proved until the very end), he then moves on to prove that

7 Given a sphere and any of the five regular solids with equal surface, the sphere is greater.

8 Of solid bodies with the same surface, the one with more faces is the greatest.

9 Given a cone or a cylinder with the same area as a sphere, the sphere will be greater.[33]

The proof of 7 appeals to Archimedes' *Sphere and Cylinder*, but it also mentions Pappus in the same breath: "this is in fact evident from the things which Archimedes proved in the sphere and cylinder and from the other lemmas submitted by me."[34] Proposition 8, rather than being proved, is, I think, taken to be evident from what is said later, when the five regular solids are compared with each other individually; in the meanwhile, Pappus states their order according to size, and reminds the reader that what happens with solids is similar to

---

[29] The proof of 5 at 348.1–350.18. Archimedes at *Sph. Cyl.* I 42 (about the area of a spherical segment less than the hemisphere) and 43 (about the area of a spherical segment greater than the hemisphere). Knorr (1978d) argues that these passages in Pappus derive from early Archimedean works.

[30] 350.10: ἐπεὶ διὰ τὸ πρὸ δύο λῆμμα.

[31] The whole passage at 352.11–358.18.

[32] 358.25 ff.; the same result in *El.* XIII 18.

[33] The three propositions at 358.30–362.16 and in Zenodorus, *De isop. fig.* 374.12–379.23.

[34] 360.18–21: τοῦτο γὰρ ἐκ τῶν ὑπ' Ἀρχιμήδους δεδειγμένων ἐν τῷ περὶ σφαίρας καὶ κυλίνδρου καὶ τῶν ἄλλων ὑφ' ἡμῶν ὑποτεταγμένων λημμάτων ἐστὶ φανερόν.

what happens with plane figures. The proof of 9 appeals to Archimedes. Pappus then shifts to yet another string of auxiliary lemmas, this time in order to attain, in a new "excursus", the formulation of the volume of a sphere:

9 Every sphere is equal to a cone whose base is the surface of the sphere and whose height is its radius.[35]

Now, this result was one of Archimedes' most famous discoveries. When Cicero, during his time as governor of Sicily, allegedly found the Syracusan mathematician's long-lost grave, he declared that it was engraved with a sphere inscribed in a cylinder, symbolizing this very theorem. Pappus is thus (here as on many other occasions in book 5) stating the obvious, something which was already well known, possibly not only to mathematicians. Indeed, he recognizes that what he says has been proved by Archimedes, and claims that he will prove it differently, after premising some lemmas, among which is a version of the lemma traditionally used in "exhaustion" procedures.[36] This existed in at least two, by and large equivalent, forms: the "Euclidean" one, based on continuous reciprocal subtraction ("If there are two unequal magnitudes and from the greater there be subtracted more than its half, from the remainder more than its half and so on, there will be left some magnitude which will be less than the lesser of the first two"), and the "Eudoxus-Archimedes" one, based on ratios ("Of unequal magnitudes the greater exceeds the lesser by such a magnitude as, when added to itself, can be made to exceed any assigned magnitude of the same kind").[37]

Pappus' lemma is closer to the Euclidean version, in that it applies specifically to arcs and tangents to the extremities of those arcs, and consequently is of use in dealing with circumferences and circumscribed polygons.[38] In fact, it is used to

---

[35] 398.11–404.8.

[36] 362.17–21: τὰ δ' ὑπὸ τοῦ Ἀρχιμήδους, ὡς εἴρηται, δειχθέντα καὶ ἄλλως ἀποδείξομεν. Cicero in *Tusculanae Disputationes* V 32.64–66. For a description of the method of exhaustion see Heath (1921).

[37] Euclid, *El.* X.1 and Archimedes, *Sph. Cyl.* I axiom 5, respectively. Cf. Hjelmslev (1950); Knorr (1978d); Vitrac (1992).

[38] 382.1–382.18.

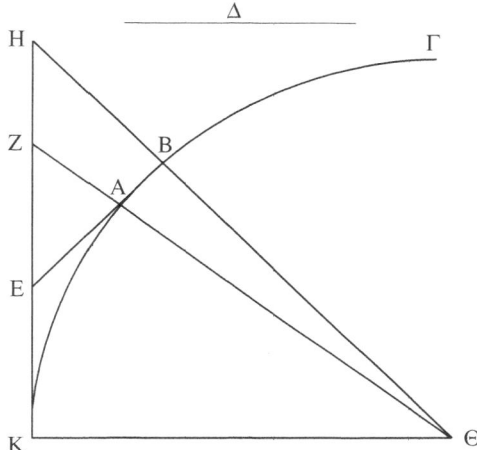

Fig. 2.1. Given a line Δ, we want to prove that it is possible to divide the arc KA *ad infinitum* to the point that, if we draw the tangents KE and AE, they will be smaller than Δ. We posit KH = Δ; we take the arc KA < KB, and it is known from previous demonstrations that AE = KE. If we prolong the radius ΘA until Z, we have that KE + AE < KZ, because, in the right-angled triangle ZAE, the hypotenuse ZE is greater than the cathetus AE, and thus greater than KE = AE. Consequently, KZ > AE + KE, and *a fortiori* KH = Δ > KE + AE.

prove the formulation of the area of a sphere, which Pappus treats as a particular case of his eleventh lemma: "Any segment of a sphere is equal in surface to a circle whose radius is the line from the vertex of the segment to its base."[39] The same result had been expressed (once again) by Archimedes in

---

[39] 382.19–386.21. Hero, *Metrica* 88.25–30 has this same proposition and says that it is taken from Archimedes' *Sphere and Cylinder*. Knorr (1978d) 240, sees both this and Pappus' proposition as taken from an early version of Archimedes' *Sphere and Cylinder*. According to Knorr, Hero would be relying on the same early version which Pappus would later also use. Also to be noted is that Hero's and Pappus' term for vertex is πόλος, while Archimedes' is κορυφή; Knorr indicates that πόλος, a term also used in Autolycus of Pythane (end of fourth century BC), would fit with the hypothesis of an early date for the source of Pappus' account. Pappus, however, regularly uses πόλος in connection with a sphere in the *Collection* (especially in book 6, devoted to astronomy) and in his commentary on Ptolemy. Κορυφή is also used, but most often about polyhedra. I remain unconvinced that Archimedes would have changed terminology between an early and a later version (if such double versions indeed existed). Moreover, Hero, at the beginning of the

three separate theorems (about the area of a sphere, of a segment of a sphere lesser or greater than a hemisphere), which in the *Collection* we find unified in a single general statement.[40] In a sort of approximation procedure to the volume of the sphere, Pappus then considers in turn all the possible variations on the theme of solids produced by the rotation of various figures: first a triangle rotating around one of its sides; then a triangle rotating around a line external to it, in three different positions; then a quadrilateral rotating around a line external to it; then a pentagon; then any polygon; then finally a polygon inscribed in a semicircle which rotates around the diameter of the semicircle.[41] The final step will be the proof of 9.

Before turning to the final part of his account, Pappus says: "and these are the things *about* what has been proved by Archimedes in the *Sphere and Cylinder*."[42] The nature of the relationship between the two texts is neither that of a commentary, nor that of an anthology or collation of results. It is not a commentary, because Pappus' reader in book 5 is not expected to have Archimedes' book at hand, and this is made quite clear; yet other extant commentaries by Pappus presuppose that the text being commented upon is available to the reader. It is not an anthology or a simple copying from Archimedes, because Archimedes' results are modified according to clearly definable criteria, and because Pappus says that this is "about" Archimedes – when elsewhere he *quotes* Archimedes, he says so.

The fourth and final string of lemmas, which this time come before the proof of the third main result, has the aim of comparing the five regular bodies with each other. The main ref-

---

Metrica 2.18 and again just before the passage in question (*ibid.* 86.30–31) quotes *verbatim* the proposition on the area of the sphere as we have it now in Archimedes' text (*Sph. Cyl.* I 33). I am inclined to attribute Hero's quotation to the obvious similarities between the two formulations, or, if must be, to an intermediate source (post-Archimedes).

[40] Archimedes, *Sph. Cyl.* I 33; 42; 43.

[41] 388.22 ff.

[42] 410.22–23: Καὶ τὰ μὲν περὶ τῶν ὑπὸ Ἀρχιμήδους δειχθέντων ἐν τῷ περὶ σφαίρας καὶ κυλίνδρου τοσαῦτ' ἐστίν. Italics mine.

erence texts are in this case book thirteen of Euclid's *Elements* and Hypsicles' *Book fourteen of the Elements*. The results that Pappus arguably draws from Hypsicles are modified by skipping some intermediary steps with the help of Theodosius' *Sphaerics*, generalizing some particular results and incorporating what in Hypsicles was a separate auxiliary lemma into the main body of the proof.[43]

It is more difficult to assess Pappus' treatment against Euclid's: his debts are undeniable and acknowledged, but the main part of *Elements* XIII, devoted to the construction of the five regular solids within the sphere, has a more evident parallel in the final part of book 3 of the *Collection*,[44] which reads like an exemplary exercise in analysis-and-synthesis constructions, aimed at inscribing figures in a sphere. We again find mention of Theodosius' *Sphaerics*[45] and particular attention is again paid to sub-cases.[46] The order of construction does not follow Euclid's: in the *Elements* we have pyramid, octahedron, cube, icosahedron, dodecahedron, while book 3 has pyramid, cube, octahedron, icosahedron, dodecahedron. Again, in Euclid the enunciation of each construction expresses the side of the polyhedron in question in terms of its relation to the radius of the circumscribed sphere. In Pappus, on the other hand, the relation between the side of a polyhedron and the diameter of the sphere is contained in the demonstration, but not emphasized in the enunciation. The focus of interest is clearly not the same: Euclid is attempting to establish the relations between different geometrical objects, Pappus to determine a construction on the basis of a given set of data. It is not incidental that Pappus' account in book 3, although it presents some significant variations from Euclid's, is also unlike the one in book 5. Above all, it does not

---

[43] 416.17–418.2, 428.5–430.6 and 438.23–444.17 of the *Collection*, respectively. The corresponding passages in Hypsicles are *El. XIV* 6.15–18, 32.10 ff. and 6.19 ff. Hypsicles and Pappus are associated in a scholion to Hero's *Geometrica* (223.27–30), with reference to a proposition which Hypsicles "put forth" and Pappus "proved" (to be identified with 416.17 ff.).

[44] 132.1–162.24.

[45] 132.18–9; 136.26; 148.8; 158.10–1.

[46] 136.10–138.26.

emphasize any comparisons, or any relation to the problem of isoperimetry, or any affiliation with Archimedes' results.

Book 3 as a whole has a polemic edge, in that it criticizes some propositions which had been sent to Pappus by an anonymous geometer. The results in question were also known to the philosopher Hierius and to people connected with him. Pappus criticizes, with varying degrees of intensity, the whole output of the anonymous geometer, and concludes his treatment with a sort of rehearsal of book XIII of the *Elements*, as if to reaffirm his own mathematical credentials after exposing other people's mistakes. The question of the five Platonic bodies was of philosophical interest; also, it was related to the topic of means and proportions, which is a sort of *Leitmotiv* for book 3, where the two mean proportionals and the ten types of ratio (arithmetic, geometrical, etc.) were also discussed.

To return to book 5 of the *Collection*, whereas in Euclid the comparison is "concentrated" in one proposition where all the bodies are put together, Pappus establishes a series of comparisons, each body with another one, in increasing, if somewhat quirky, order. It would normally be pyramid, cube, octahedron, dodecahedron, icosahedron, but the propositions in Pappus are as follows:

A cube is greater than a pyramid.
An octahedron is greater than a cube.
An icosahedron is greater than an octahedron.
An icosahedron is greater than a dodecahedron.
A dodecahedron is greater than an octahedron.

Compared to Euclid, Pappus in book 5 does not show the concern for the study of incommensurable magnitudes that many critics have seen in the *Elements*, book thirteen, or attributed to its sources.[47] Neither does his chief aim seem to be the determination of volumes and areas, evident in Archimedes. As for Hypsicles and Zenodorus, their strict reliance on Euclid and Archimedes, respectively, and the fragmentary state of their extant works makes it difficult for us to assess

---

[47] Sachs (1917); Mueller (1981).

whether Pappus was just deploying some of their results, reworked as they might be, or also reproducing a more general outlook. On the other hand, some features emerge which in the course of our analysis of the *Collection* will be revealed to be typical. For instance, there is the recourse to Theodosius to "speed up" some results handed down from the tradition. Then again, there is the strategic way Pappus deploys his sources: he underlines his debts to Euclid and Archimedes in order to point out that he has access to, and fully understands, these sources; he underlines his difference from Archimedes in order to point out that he can produce mathematics comparable, or at least complementary, to these sources. As for his lesser known predecessors, he modifies them along distinctive lines, sometimes making explicit the conditions necessary for the solution (usually to a problem); sometimes examining subcases of a proposition. These two things are somehow connected in that they both imply a manipulation of the geometrical object or objects, or, in other words, they imply playing with some elements of the construction. Another notable thing, which is more characteristic of book 5 than of the rest of the *Collection*, is the way in which Pappus sometimes changes the order of the propositions as they are found in his sources and postpones the demonstration of lemmas to the demonstration of the result they are used to prove. This is not very common in classical mathematical texts such as, for instance, the *Elements*. R. Netz has studied argumentative structures in, among others, Aristotle, Euclid and Archimedes, and argues that an arrangement of results where auxiliary propositions are proved *after* they are used to prove something else is meant to emphasize the something else, the main result, rather than the argumentative structure itself.[48] In other words, on this interpretation, Pappus' main interest would be in communicating the main results (including the main results of the excursus), while the details of their demonstrations take second place.

---

[48] Netz (forthcoming b). Thus, Aristotle emphasizes the main results, whereas Euclid and Archimedes put more emphasis on the demonstrative process used to get to those main results.

From a mathematical point of view, then, in book 5 Pappus takes some previous results, modifies them along characteristic lines, and marks out debts or claims to variations with respect to the past strategically, in order to produce an account that is meant to convey both his mathematical skill and his knowledge of the tradition.

But there is more to book 5 than mathematics. The cluster of problems connected with isoperimetry and especially with the five Platonic bodies had a long and extremely rich history in at least two fields and lines of tradition. In philosophy, the theory of the five bodies, in its various NeoPythagorean and NeoPlatonist versions, represented a corner-stone for Platonist physical theories. In my opinion, Pappus was writing for an audience which was more familiar with this second tradition than with the mathematical one. We have various signals that his intended recipient Megethion, whether real or fictional, was not a geometer by profession, or rather that he had no particular mathematical abilities.

First of all, there are the numerous references, both direct and implicit, to Archimedes. Along with Euclid, who, as it happens, is the other person to be picked out for special mention in Pappus' account, the Syracusan geometer was one of the few people to have gained fame and fortune in extra-mathematical circles – a big name that, unlike Zenodorus or Hypsicles, would have been recognized by a wider public. Anecdotes about Archimedes and, to a lesser extent, Euclid, found their way into historical works and philosophical and moral treatises, including Platonist literature.[49] Popular knowledge may have included being aware that Archimedes had made discoveries about the sphere and the cylinder,[50] yet a layman (and here we can include educated people with no particular interest in geometry) would not have been expected actually to have read the treatise by the same title.

We can get a better idea of the audience of book 5 by

---

[49] E.g. Plutarch, *Vita Marcelli* 14ff. for Archimedes; Proclus, *In Eucl.* 41.5–8; 63.18–64.2 for Archimedes; 68.13–17 for Euclid; Stobaeus, *Anthologia* 205.15–19 (*apud* Johannes Damascenus) for Euclid.

[50] Cf. e.g. Cicero, *Tusc. Disp.* V 32.64–66.

comparison with other parts of the *Collection*. The Pandrosion and Hermodorus addressed elsewhere are, respectively, a teacher of mathematics and someone who has access to the rather advanced *Treasure of Analysis*.[51] In book 6 Pappus assumes that his reader is familiar with authors such as Theodosius, Autolycus, Aristarchus (these last two must have been little read and probably difficult to come by in the fourth century AD); that his reader knows that Euclid's *Phenomena* is incomplete and that he has access to some of Ptolemy's books.[52] In book 4 Pappus tells his addressee that he is going to give him some tips for a better understanding of Archimedes' *Spirals*. The tips amount to a rather complex extension – not a paraphrase of the text, nor even a commentary on it.[53] On the contrary, here in book 5, Megethion is assured that he need not bother to seek out Archimedes' book: Pappus himself can save him the trouble and produce another version of the proof in question i.e. 3, the formulation of the area of the circle.[54]

It is remarkable, I think, that in the case of book 4 the reader is left to fend for himself through a text difficult to understand even for skilled mathematicians – indeed, a text that was probably only known to experts – while, in the case of book 5, the reader is not even expected to be acquainted with one of Archimedes' most famous discoveries, let alone with Archimedes' text or with the general lines of the procedure. These marked double standards in the case of an author usually severe towards incompetence can only be explained on the basis of two different kinds of intended readership.

Again, Pappus introduces his comparison of the five Platonic bodies thus: "[I will write the comparison of the five

---

[51] Pandrosion at 30.2; Hermodorus at 634.1 and 1022.1 (books 3, 7 and 8 respectively).

[52] 632.18–19 and 632.20–22, respectively.

[53] 298.3ff.

[54] 312.25–314.3: Πρὸς τὸ μὴ δεῖσθαι τοῦ Ἀρχιμηδείου συντάγματος ἕνεκεν μόνου τοῦ θεωρήματος τούτου. In the commentary on Ptolemy (253.6), Pappus again declares that there is no need to peruse Archimedes' book about the theorem for the area of the circle, because he has provided it already in the scholia to book 1. (He is probably referring to Book 1 of the same commentary, now lost). See this in Knorr (1978c).

bodies] not by means of the so-called analytical investigation, by means of which some of the ancients produced the proofs ... but by means of the synthetic method set up by me in order to achieve the clearest and most concise form."[55] Apart from a desire to differentiate himself from previous mathematicians which we have already noticed, Pappus' emphasis on a clear and concise mode can be seen as being in line with his emphasis on the main results, rather than on the argumentative process itself. A reader with a more pronounced mathematical flair would perhaps have been expected to benefit from the analytic account, and elsewhere Pappus has no qualms in inflicting on his audience lengthy analyses even of secondary lemmas.[56] In fact, his own treatment of the five solids in book 3, centered around the construction of each regular polyhedron within the sphere, is arranged as a string of analyses followed by shorter syntheses, but then, as we have said, book 3 is addressed to a teacher of mathematics.

The fact that in book 5 Pappus is interested in getting through the main results first of all, unhindered, so to speak, by the demands of the deductive order, and also the fact that he repeatedly states those main results (at the beginning of the chain of deduction, at the end of the lemmas before the proof, and sometimes again at the end of the proof) – all this combines to give the impression that he is trying to hammer them into the reader's head in a way far removed from his own practice elsewhere. Take, as an example, the comparison of the five bodies: Euclid, as we have said, concentrates everything in a single proposition. Pappus, instead, states the order of the Platonic bodies a first time just before proving the second main result. He adds that, by analogy with plane figures, the more faces a polyhedron has, the greater its volume will

---

[55] 410.27–412.3: οὐ διὰ τῆς ἀναλυτικῆς λεγομένης θεωρίας, δι᾽ ἧς ἔνιοι τῶν παλαιῶν ἐποιοῦντο τὰς ἀποδείξεις ... ἀλλὰ διὰ τῆς κατὰ σύνθεσιν ἀγωγῆς ἐπὶ τὸ σαφέστερον καὶ συντομώτερον ὑπ᾽ ἐμοῦ διεσκευασμένας.

[56] E.g. 276.32–280.19 (in book 4). According to Marinus, *In Euclidis Data* 256.22–25, one should follow an analytical, not a synthetical mode of discourse for teaching, "as Pappus sufficiently proves". This could be taken as supporting evidence for the fact that the addressee of book 5 was not a student of mathematics.

be. In a different context this would probably have sufficed –
yet Pappus repeats the sequence again at the beginning of the
string of comparisons: we are again told that the icosahedron
is the greatest, then the dodecahedron, the octahedron, the
cube, down to the pyramid.[57] To conclude we are reminded
that the polyhedron with more faces is greater than those with
equal surface but fewer faces.

In book 5 Pappus uses both cross-references and references
to other authors in an extremely detailed manner. That is, he
quotes not just "as in the *Elements*", but "as in book x [and
sometimes proposition y] of the *Elements*." Also, his cross-
references for the most part indicate exactly what position the
lemma has in the treatise, e.g. two propositions before.[58]
Pappus' numbering often does not correspond with our extant
text of the *Sphere and Cylinder* or the *Elements* or, more surprisingly, even with our extant text of the *Collection*. In any
case, such a practice of "footnoting" is very unusual, as one
can see by the sheer number of instances in this book when
compared with the other parts of the *Collection*. I take it as
further evidence that book 5 was meant for an audience who
would have missed the significant appeals to Euclid and Archimedes, and needed to be reminded that what Pappus was
doing had a long and prestigious history. Name-dropping, in
antiquity (there are countless examples from, for instance,
grammar or philosophy) as in modern times, serves the purpose of stressing that the author is knowledgeable and possesses a direct line to the tradition; this was an empowering
gesture, especially at a time when access to old texts was
rather limited. Precise quotation, which could signify that the
person quoting knew his classics by heart (again, witness parallels in other fields, such as the frequency of citations from
memory of Homer), denoted an even higher degree of expertise and was bound to impress the non-*cognoscenti*.

---

[57] 452.13–18.
[58] Cross-references at 332.25; 334.2,7,11; 350.10 and *passim*; "specific" references at
314.9; 378.8; 380.14 and *passim*.

2.2 Creations

The five Platonic bodies owe their name to the fact that they are found in the *Timaeus*, a dialogue which describes, among other things, how the universe and all its components were fashioned by the Demiurge. In Plato's account, each of the four traditional elements had a well-determined geometrical structure. They were made up by triangles following a regular pattern which produced in turn the regular solid bodies (cube for earth, pyramid for fire, octahedron for air, icosahedron for water) with the dodecahedron left as "decoration."[59] The text also praised the sphere as all-encompassing and beautiful, though that need hardly reflect any knowledge of isoperimetric theorems.[60]

Plato's motives and the role of the mathematical structuring of the world-system in his philosophy are beyond the scope of my study: what interests us here is more the way in which his account was seen and used by people contemporary to, or slightly earlier than, Pappus. The purely mathematical content of the *Timaeus* is very vague: the only solids referred to by name are cube and pyramid, and it is only from the details of their construction that we infer that the third, fourth and fifth solid are octahedron, dodecahedron and icosahedron. Also, nowhere does Plato maintain that there are *only* five such bodies, even though his otherwise superfluous addition of the fifth body can only be understood if there was already some idea that the regular solids came as a group of five.[61]

---

[59] Plato, *Timaeus* 55c.
[60] *op. cit.* 33b: "[W]herefore He wrought it into a round, in the shape of a sphere, equidistant in all directions from the centre to the extremities, which of all shapes is the most perfect and the most self-similar."
[61] Sachs (1917), 48, 92, claims that "Plato knew Euclid's conclusion that contains the proof that there are only five regular bodies" (on the basis of the passage at *Tim.* 54e–55a: "And when four equilateral triangles are combined so that three plane angles meet in a point, they form one solid angle ... And the second solid is formed from the same triangles, but constructed out of eight equilateral triangles, which produce one solid angle out of four planes ...") and attributes the theorem that there can only be five regular solids to Theaetetus, 92.

After Plato, testimonies on the issue are found in a wide variety of authors. In a passage from *On the heavens* Aristotle observes that of all motions "proceeding from the same point to the same point, the circle is least," so that, swifter motions corresponding to lesser paths, the heaven "must be spherical in shape."[62] Simplicius' comment on this passage is that isoperimetric theorems must have been proved before Aristotle, even though, he remarks, they were demonstrated more generally by Archimedes and Zenodorus.[63] Ptolemy also mentions, as something already known to the reader, the notion that the heavens, being spherical, are greater than all other bodies, because, of different figures with equal perimeter, those with more angles are greater, therefore circle and sphere are greater than all other plane and solid figures, respectively.[64]

Quintilian, in a passage to which we have already drawn attention in chapter one, mentions the problem of isoperimetry as the perfect example of a case where geometry can help the orator show off his knowledge (so to speak). It is worth quoting at length:

Who is there who would not accept the following proposition? "When the lines bounding two figures have the same measure, it is necessary that also the spaces contained within those lines be equal." But this is false, for much more depends on what is the shape of that contour, and those historians have been reprimanded by geometers who believe that the time taken to circumnavigate an island is a sufficient indication of its size. For the more perfect the figure, the more capacious it is.

Quintilian then adduces some examples that seem to be taken from land-surveying, since they mention the Roman acre, and concludes: "it further follows that it is perfectly possible for the space enclosed to be less, though the perimeter be greater. ... for even one who is inexpert can see that in hills or valleys the extent of ground enclosed is greater than the sky

---

[62] Aristotle, *De caelo* 287a28–31: ἀλλὰ μὴν τῶν ἀφ' αὐτοῦ ἐφ' αὐτὸ ἐλαχίστη ἐστὶν ἡ τοῦ κύκλου γραμμή ... ὥστ' εἰ ὁ οὐρανὸς ... σφαιροειδῆ αὐτὸν ἀνάγκη εἶναι. Cited by Knorr (1986a), 273.

[63] Simplicius, *In Aristotelis De Caelo* 412. Zenodorus is also mentioned by Proclus, *In Eucl.* 165.24.

[64] Ptolemy, *Syntaxis* I 13.10–20.

2 BEES AND PHILOSOPHERS

over it."[65] Quintilian suggests that notions about isoperimetry, as an interesting paradox about the relation between contour and area, had entered general knowledge.[66]

But the issue took still another guise at the hands of Plato's "legitimate" successors. The *Timaeus* was central to the Neo-Platonist doctrine of mathematics as the mediator between the world of intelligibles and that of physical beings. In this context, the discipline acquired crucial importance, both as knowledge in its own right and as a model of discourse for describing higher entities by analogy. The validity of mathematical knowledge itself was warranted by the fact that the universe had an underlying geometrical structure, expressed by the five bodies and the all-encompassing sphere. Neo-Platonist and NeoPythagorean traditions contain a number of reports which attribute the doctrine and discovery of the five bodies to Pythagoreans or emphasize the "Pythagoreanism" of the *Timaeus*.[67]

The earliest extant (for us and very likely for them in the third or fourth century AD) systematic account on the subject, however, is to be found with Euclid. It is intriguing, then, that Proclus should feel compelled to read the whole of Euclid's

[65] *Institutio oratoria* I 10.39–45: "Nam quis non ita proponenti credat? Quorum locorum extremae lineae eandem mensuram colligunt, eorum spatium quoque, quod iis lineis continetur, par sit necesse est. At id falsum est. Nam plurimum refert, cuius sit formae ille circuitus; reprehensique a geometris sunt historici, qui magnitudinem insularum satis significari navigationis ambitu crediderunt. Nam ut quaeque forma perfectissima ita capacissima est.... Ergo etiam id fieri potest, ut maiore circuitu minor loci amplitudo claudatur.... Nam in collibus vallibusque etiam imperito patet plus soli esse quam caeli." (Loeb trans. with my modifications.)

[66] A wide diffusion was certainly enjoyed by the legend according to which Dido had cunningly managed to obtain the land on which Carthage was founded. She had been accorded as much land as could be covered by a bull's hide, but cut the hide into very thin strips, with which she surrounded a huge stretch of territory (again, an unexpected relation between contour and area). Vergil only hints at the episode, *Aeneid* 1.368. The sources are collected in Pease (1967), 16–17; see also Müller (1953).

[67] Sachs (1917), 9–15 has a catalogue of sources mentioning the five regular bodies in connection with Pythagoras or the Pythagoreans (Plato is sometimes included among those latter): Speusippus (*apud* the pseudo-Iamblichus); Theophrastus (*apud* Achilles and Aetius, who even adds: "Indeed Plato in these things pythagorizes" (Πλάτων δὲ καὶ ἐν τούτοις πυθαγορίζει) in DK 28A44); Iamblichus; pseudo-Iamblichus; Hermias; Proclus. See also Burkert (1962); Donini (1982); O'Meara (1989); Zhmud (1997).

Elements as essentially aimed at producing the results of book thirteen: "Euclid belonged to the persuasion of Plato and was at home in this philosophy; and this is why he thought the goal [or end] of the *Elements* as a whole to be the construction of the so-called Platonic figures."⁶⁸ That the *Timaeus* is relevant in this respect is, again in Proclus' words, evident. He summarizes the aim of book I of the *Elements* thus: "[to present] the triangle and the parallelogram. For these are the genera that include the causal principles of the elements ... the triangle and the square, from which the figures of the four elements are constructed.... Consequently the aim of the first book [is to contribute] to the full understanding of the cosmic elements."⁶⁹

We have seen in the previous chapter that NeoPytha-Platonist attitudes towards Euclid included Iamblichus' harshly-termed criticisms (Euclid is accused of manifest mistakes, ignorance and confusion). It is important not to assume that all late ancient philosophers thought the same about mathematics.⁷⁰ The range of attitudes to Euclid is paralleled in some representations of other mathematicians who were not traditionally associated with Pythagoreanism: Apollonius and, above all, Archimedes. Not only was this latter hugely famous as a mathematician, but some facets of his persona, particularly his reputation as the Plutarchean sage who despised practical applications and only cared about abstract

---

⁶⁸ Proclus, *In Eucl.* 68.20–3: καὶ τῇ προαιρέσει δὲ Πλατωνικός ἐστι καὶ τῇ φιλοσοφίᾳ ταύτῃ οἰκεῖος, ὅθεν δὴ καὶ τῆς συμπάσης στοιχειώσεως τέλος προεστήσατο τὴν τῶν καλουμένων Πλατωνικῶν σχημάτων σύστασιν; cf. also 70.19–71 5: "looking at its subject-matter, we assert that the whole of the geometer's discourse is obviously concerned with the cosmic figures."

⁶⁹ *op. cit.* 82.13–83.2: ἐν τούτῳ δὴ οὖν τῷ βιβλίῳ τὰ πρώτιστα καὶ ἀρχοειδέστατα σχήματα τῶν εὐθυγράμμων παραδίδοται, τό τε τρίγωνον λέγω καὶ τὸ παραλληλόγραμμον· ἐν γὰρ τούτοις ὡς ἐν γένει περιέχεται καὶ τὰ αἴτια τῶν στοιχείων, τό ... τρίγωνον καὶ τετράγωνον, ἀφ' ὧν τὰ σχήματα τῶν τεττάρων στοιχείων ἔσχεν τὴν σύστασιν.... συνήρτηται δὴ οὖν ὁ σκοπὸς τοῦ πρώτου βιβλίου πάσῃ τῇ πραγματείᾳ καὶ συντελεῖ πρὸς ὅλην τὴν τῶν κοσμικῶν στοιχείων θεωρίαν. He also comments that one of the consequences of a certain theorem will be to "prepare us for the teaching of the *Timaeus*", *ibid.* 384.2–4.

⁷⁰ Iamblichus, *In Nicom.* 20.10–14; 23.18 ff; 25.24; 30.28; 74.23 ff. and cf. Mueller (1987b). Note that when Pappus (as we shall see in chapter four) criticizes Archimedes and Apollonius, his tone is more moderate: "a not small mistake," 270.28 ff.

speculation, fitted in very well with Platonist ideas. Nonetheless, no direct links were drawn between Archimedes and the Pythagoreans or Archimedes and Plato. Iamblichus' attitude (as quoted by Simplicius) is then particularly significant: he attributes the formulation of the area of the circle to a "Sextus the Pythagorean" and limits Archimedes' contribution, usually recognized to be major, to some investigations by means of, maybe, spiral lines (the text presents a lacuna). Simplicius himself expresses some misgiving about the value of this report:

Aristotle did not know this [the formulation for the area of the circle] at all, it seems; but it was discovered by the Pythagoreans, says Iamblichus, as is clear from the proofs of Sextus the Pythagorean, who at the beginning learnt the method of the proofs according to the tradition. And afterwards, he says, Archimedes by means of ... [The names of other mathematicians follow: Nicomedes, Apollonius, Carpus] ... as Iamblichus reports. And it is most extraordinary that the most learned Porphyrius omits this – in fact he seems to say that there is a proof, according to which there is a figure of a square that approximates the circle, as for the other figures, but it is not transmitted and not yet even discovered.[71]

It is also intriguing that in Proclus' commentary on the *Elements* Apollonius should be regularly pitted against the Pythagoreans, Aristotle or Euclid, and should regularly come out of the comparison in a bad light: he made false statements about curves; he messed up the definition of angles; he even tried to demonstrate an axiom, and inevitably failed. Compared with Apollonius' construction of a certain problem, Euclid's is "much better, since it is simpler and proceeds from the principles."[72]

---

[71] Simplicius, *In Categorias* VII 192.15–25 (the relative passage in Aristotle is 7b15): τοῦτο δὲ 'Αριστοτέλης μὲν, ὡς ἔοικεν, οὔπω ἐγνώκει, παρὰ δὲ τοῖς Πυθαγορείοις ηὑρῆσθαί φησιν 'Ιάμβλιχος, ὡς δῆλόν ἐστιν ἀπὸ τῶν Σέξτου τοῦ Πυθαγορείου ἀποδείξεων, ὃς ἄνωθεν κατὰ διαδοχὴν παρέλαβεν τὴν μέθοδον τῆς ἀποδείξεως. καὶ ὕστερον δέ, φησίν, 'Αρχιμήδης διὰ τῆς † ... ὡς 'Ιάμβλιχος ἱστορεῖ. καὶ θαυμαστὸν ὅτι τοῦτο τὸν πολυμαθέστατον ἔλαθεν Πορφύριον, ὃς φαίνεται μέν, φησίν, ὅτι ἔστιν τις ἀπόδειξις, καθὸ ἔστιν σχῆμα τετράγωνον κύκλῳ παραβαλεῖν ὥσπερ καὶ ἄλλα σχήματα, κατείληπται δὲ οὐδέπω οὐδὲ εὕρηται.

[72] Proclus, *In Eucl.* 105.13 ff.; 123.14 ff.; 183.13 ff.; 194.9 ff.; 280.9–11, respectively. Note that, although Apollonius gets criticized by Pappus in the *Collection* (as we will see in chapters four and five), in the *Commentary on book 10* he is hailed as a genius, 63.

There is a sense then in which philosophers within Neo-Pythagorean and NeoPlatonist traditions had to re-write or opportunely construct previous mathematical traditions in order to preserve both the importance of mathematics within their philosophy and the importance of themselves as *the* authorized depositories of knowledge – a knowledge which, crucially, included mathematics. Proclus' attempt at "annexing" Euclid to his own tradition and Iamblichus' attempt at pushing Euclid aside to make space for his own people represent different solutions to what can be seen as the same problem: the problem of coming to terms with competing traditions, which shared the same issues but approached them in different ways. I take it that in book 5 of the *Collection* Pappus is in his turn coming to terms with the problem.

2.3 Boundaries

The philosophers say that the first god, maker of all things, has opportunely given to the cosmos the shape of a sphere, choosing the most beautiful among things that are, and declare the natural properties inherent to the sphere and add that the sphere is the greatest among solid figures with the same area. They also say other things inherent to the sphere which are more than clear and do not require great persuasion, whereas that the sphere is greater than the other figures the philosophers do not prove, but only assert, and it is not easy to be persuaded of it without further investigation.[73]

This passage comes immediately after the end of the second excursus, before the treatment of the second main result and immediately before the description of Archimedes' thirteen semi-regular solids. That the philosophers Pappus refers to are Platonists seems clear from the fact that they would have been the obvious candidates at the time, given their decided interest

---

[73] 350.20 ff.: Τὸν πρῶτον καὶ δημιουργὸν τῶν πάντων θεὸν οἱ φιλόσοφοί φασιν εἰκότως τῷ κόσμῳ σχῆμα περιθεῖναι σφαιρικὸν ἐκλεξάμενον τῶν ὄντων τὸ κάλλιστον, τά τε προσόντα τῇ σφαίρᾳ φυσικὰ συμπτώματα λέγοντες ἔτι καὶ τοῦτο προστιθέασιν ὅτι πάντων τῶν στερεῶν σχημάτων τῶν ἴσην ἐχόντων τὴν ἐπιφάνειαν μεγίστη ἐστὶν ἡ σφαῖρα. τἄλλα μὲν οὖν ὅσα προσεῖναι λέγουσιν αὐτῇ πρόδηλά τέ ἐστιν καὶ παραμυθίας ἐλάσσονος δεῖται, τὸ δ᾽ ὅτι μείζων ἐστὶ τῶν ἐλλων σχημάτων οὔθ᾽ οἱ φιλόσοφοι δεικνύουσιν, ἀλλ᾽ ἀποφαίνονται μόνον, οὔτε παραμυθήσασθαι ῥᾴδιον ἄνευ θεωρίας πλείονος.

in the *Timaeus*. The most likely guess could be Iamblichus, who, as we have seen, talks a lot about mathematics in a way which Pappus may have not entirely approved of. His commentary on Nicomachus mentions the circle as something which "contains and encloses any figure of plane polygon, and the sphere [any figure] of solid [polyhedron]." Iamblichus' commentary on the *Timaeus* is now lost, but fragments of it are contained in Proclus' commentary on the same text and it seems that he talked at length about the properties of the sphere as the only figure capable of containing all the elements, unlike the other geometrical figures, which have a multiplicity of surfaces and angles.[74]

Proclus' commentary on the *Timaeus* draws on Iamblichus also for a number of "philosophical" proofs of a statement by Plato to the effect that the universe is spherical; to the philosophical and physical arguments (these latter taken from Aristotle) Proclus adds "mathematical" ones: various remarks to do with astronomy (for instance, the variation in the relative lengths of day and night) and a concise report of "other people" 's investigations. These unnamed authors, using results by Euclid and Archimedes, had first proved that the polygon with a greater number of angles is greater than other regular polygons with the same perimeter, then that the circle is greater than plane figures with the same perimeter, and finally that the sphere is greater than other solids with the same surface, and especially greater than the five regular so-called Platonic solids. Proclus declines to give more details, adding that he will append a collection (συναγωγή) of mathematical theorems to his text for those readers who are curious about this particular aspect of the *Timaeus*. This collection is now lost.[75]

Could it be that one of the unnamed authors in question was Pappus? Proclus knew Pappus: he mentions him explicitly in the commentary on Euclid, and Proclus' pupil Marinus also

---

[74] Iamblichus, *In Nicom.* 61.9 ff. (κύκλος χωρητικός ἐστι καὶ περικλειστικὸς παντὸς πολυγώνου ἐπιπέδου σχήματος καὶ ἡ σφαῖρα στερεοῦ); Iamblichus *ap.* Proclus, *In Timaeum* 3.72.6, respectively.

[75] Proclus, *In Timaeum* 3.75.18–76.29.

mentions Pappus in his commentary on Euclid's *Data*.[76] In fact, it could be argued that some passages in Proclus are a reaction to Pappus: in the commentary on the *Timaeus*, Proclus declares that he will append Archytas' solution to the duplication of the cube at the end of the treatise (probably in the same "collection" mentioned above), but he will not include Menaechmus' or Eratosthenes' solutions to the same problem, the first because it involves conics and the second because it employs a moving ruler. Now, in *his* treatment of the duplication of the cube Pappus, as we shall see in chapter four, not only unproblematically reports Eratosthenes' solution (on my interpretation, he even selects it from the material available to him), but he employs a ruler for his own solution and praises the use of mechanical instruments in geometry at some length.[77] Again, in the commentary on Euclid, Proclus dismisses the so-called "linear" curves as an object for discussion – whereas a prominent part of Pappus' book 4 is devoted to a careful and favourable account of linear curves.[78] And, in a curious parallel to the introduction to book 5 of the *Collection*, Proclus' commentary on Euclid contains a passage on the mathematical knowledge of animals. About the theorem that in any triangle the sum of two sides taken is greater than the remaining side, the Epicureans say:

it is evident even to an ass and needs no proof; it is as much the mark of an ignorant man, they say, to require persuasion of evident truths as to believe what is obscure without question.... That the present theorem is known to an ass they make out from the observation that, if straw is placed at one extremity of the sides, an ass in quest of provender will make his way along the one side and not by way of the two others. To this it should be replied that, granting the theorem is evident to sense-perception, it is still not clear for scientific thought. Many things have this character, for example, that fire warms. This is clear to perception, but it is the task of science to find out how it warms ... So with respect to a triangle let it be evident to perception that two sides are greater than the third; but how this comes about it is the function of knowledge to say.[79]

---

[76] Proclus, *In Eucl.* 189.12; 197.6; 249.20–250.12; the paradoxical theorem at 326.24 ff. could be lifted from book 3, 104.14 ff.; Marinus, *In Euclidis Data* 256.22–25.

[77] Proclus, *In Timaeum* 3.33.30–34.4.

[78] Proclus, *In Eucl.* 113.15 ff.

[79] *op. cit.* 322.4–323.3.

In sum, it could well be that, with his foray into mathematical territory, which included, as we have mentioned, an attempt at annexing Euclid, Proclus was also reacting to Pappus' stance on mathematics, especially as Pappus, in fact, is maintaining that the very operation Proclus has embarked upon was not within the domain of competence of philosophers.

That said, book 5 need not have been directed *specifically* against Iamblichus or his followers – it could have been aimed at a straw adversary, who embodied trends current at the time. Pappus was acquainted with philosophical works: in his own commentary on Euclid he mentions the *Theaetetus*, the *Laws* and the *Parmenides*;[80] again, in book 3 he unites in the same discourse on ratios Nicomachus "the Pythagorean" and some theories taken from the *Timaeus*,[81] yet in both cases there is no trace of criticism. Book 5 need not amount to evidence of a particular philosophical stance on Pappus' part: he is simply defending his own turf. In other words, Pappus might very well have subscribed to some version of the current philosophies – if indeed there was a contrast between philosophers and mathematicians, it was certainly more a matter of relative status and authority than a matter of doctrinal differences. The contention was chiefly about the role of mathematics and who was entitled to dispense knowledge of it; it was about a boundary being drawn.

It is worth emphasizing that at the time there was fierce competition between teachers. Episodes are reported of students kidnapped to secure their custom, or of violent fights between pupils of different schools.[82] At the same time, this did not necessarily mean (indeed, it rarely meant) that the contention was about different doctrines. It was more a question of securing the best reputation and the best pupils, who brought rewards in the form of fees and especially of useful connections.[83]

---

[80] Pappus, *Commentary on book 10 of Euclid's Elements* 63, 64, 72, 75, *passim*.
[81] 84.1–88.4.
[82] Jones (1964), II 1002 quotes examples from Libanius, *Orationes* 1.19–21; Eunapius, *Vitae* 483, 485; Gregorius Nazanzienus, *Orationes* 43.15–16.
[83] Evidence quoted in Mazzarino (1951); Kaster (1988), 125.

We have some pronouncements on the subject of who should be teaching what, although they are later than Pappus' time. Proclus, in concluding his commentary to the first book of Euclid's *Elements*, expresses the wish that the authors of commentaries on other parts of the text may be "aiming always at what is important and can be clearly divided, since the commentaries now in circulation contain great and manifold confusion and contribute nothing to the exposition of causes, to dialectical judgement, or to philosophical understanding."[84] Is this a stab at more "mathematical" commentaries on Euclid, such as Hero's or Pappus'? In fact, what survives of Pappus' commentary on book X of the *Elements* seems to have quite a lot to say about philosophy, but devotes its second part almost entirely to mathematics and, in any case, is a very different operation from Proclus'. Pappus distances himself from Pythagorean lore, indicates Theaetetus, Apollonius and Eudemus as his reference points, and declares that his aim is to consolidate the good that there is in Euclid, rather than attaching some extra meaning to the text.[85] We also know that at the beginning of the sixth century Damascius contrasted his own teacher Isidorus with Hypatia of Alexandria. He said: "There was a great difference between them, not only in that which distinguishes a man from a woman, but still more in that which distinguishes a philosopher from a geometer."[86] Sexist remarks apart, it is to be noted that Hypatia was well known to have taught philosophy (Platonism *et alia*) as well as mathematics.[87] What these examples suggest is that, three centuries after Pappus, the issue of what were the respective fields of competence for

---

[84] Proclus, *In Eucl.* 432.14–19.

[85] Pappus, *Comm. on book 10* 63, 64. Sachs (1917), 74–5, comments that in this work "Pappus ... gives a report on the history of the doctrine of the irrationals which is entirely sober and totally free from Pythagorean mysticism.' Indeed, Pappus exposes the story according to which the person who had revealed the incommensurability of the side and diagonal of a square was drowned as a punishment, *Comm. on book 10* 64.

[86] Damascius, *Life of Isidorus, ap.* Photius, *Bibliotheca* 242, 346: Ὁ Ἰσίδωρος πολὺ διαφέρων ἦν τῆς Ὑπατίας, οὐ μόνον οἶα γυναικὸς ἀνήρ, ἀλλὰ καὶ οἶα γεωμετρικῆς τῷ ὄντι φιλόσοφος. (Trans. Marrou (1963), 135, with my modifications.)

[87] Tannery (1880b), Marrou (1963).

philosophers and mathematicians was not settled, and that, in any case, boundaries could be opportunistically drawn, depending less on actual separations than on points of view and specific interests.

To return to Pappus, it is clear what side of the culture wars he is on: the philosophers proclaim some properties of the sphere which are so obvious (Pappus comments) that they really need no demonstration. Indeed, it is hardly necessary to persuade anyone of their truth, so evident are they. On the other hand, when it comes to a more substantial statement, the philosophers do not *prove*, but just *maintain* it, whereas some investigation is required in order to justify their claims – investigation which Pappus graciously takes upon himself. He shows from the start that he can outdo the philosophers, by adding to the five Platonic bodies the thirteen Archimedean ones, described in detail. He manages in one blow to show that they cannot discriminate between useful and obvious knowledge (to put it in extreme terms, they proclaim as earth-shattering revelations things that everybody knows), that they are not able to support satisfactorily their more interesting claims, and, finally, that they do not even know the full story, since they ignore the existence of Archimedes' results, which are introduced thus: "[there are] *not only* the five figures in the divine Plato, . . . , *but also* those discovered by Archimedes."[88]

In the light of this, it is not accidental that Euclid and Archimedes are quoted so very often, nor is it incidental that, on more than one occasion, Pappus interposes himself, as it were, between his source and the reader. When he remarks that there is no need for Megethion to peruse Archimedes' book himself, Pappus is in fact blocking direct access to the source, and acting as the only proper intermediary – in a curious parallel to the role some Platonists proposed for themselves, as filters between truth and the pupil. This attitude combines overarching authority with the desire to make things as easy

[88] 352.10 ff.: οὐ μόνον τὰ παρὰ τῷ θειοτάτῳ Πλάτωνι πέντε σχήματα . . . , ἀλλὰ καὶ τὰ ὑπὸ Ἀρχιμήδους εὑρεθέντα. Italics mine.

as possible: Pappus spares the reader any trouble, yet at the same time, as a necessary agent, as a specialist who takes care of things, he makes his own action indispensable.

The most obvious, and, for the reasons we have seen, the most effective blocking of access to a source is Pappus' treatment of Archimedes, but he is not the only case. The same thing happens again with a more generic referent, "the ancients."[89] They have provided the analytic investigation, as well as the demonstrations, of the five solids, but Pappus, as we have noted, chooses to omit it and supplies only the synthetic counterpart. Since this is more concise, it is a means of presenting the results in a more straightforward manner; the reader, however, is barred from any glimpse of the heuristic process behind them. Pappus' preference for blending the food for the reader, instead of having him chew it long and hard, as it were, indicates once again that he is addressing not a student of mathematics, who would be taught to repeat, reproduce the results and, with any luck, find some more by the same token, but a lay reader, who has to be persuaded of the result achieved, yet is not shown the machinery behind it.

The carefully arranged structure of book 5 serves more than one purpose: while on the one hand it facilitates a non-expert's approach to mathematics, on the other hand it monitors that approach, enabling the author to maintain a massive degree of control. It is as though Pappus intended to make his own role as significant as possible. This kind of author-reader relationship tends to empower the author to a greater degree than one where the reader is more at home with the subject, because from the deliberately oversimplified account the author emerges not just as a master, but as a master whose knowledge cannot be shared to its full extent.

And that the extent of this knowledge was vast we are meant to apprehend by the opening passage of book 5, where Pappus not only deploys his own professional weapons to show that, when talking mathematics, philosophers are sadly out of their depth, but is also ready to take them on at their

[89] 410.22–412.7.

own game, engaging in an epistemological debate. In a style which won Hultsch's praise,[90] Pappus presents the difference between men and animals as eminently epistemic. He tells Megethion that the god (the Demiurge?) endowed men with knowledge and with mathematics: however, even some un-reasoning animals, namely bees, possess sophisticated geo-metrical notions and some kind of virtue. Bees, in fact, are a familiar paragon of obedience, industriousness and political qualities; moreover, they are able to maximize the space in their beehives by building regular hexagonal cells. What, then, is the difference between men and bees? On the whole, only the fact that men can provide a proof of how hexagonal cells maximize space, and, above all, the fact that they are *interested* in providing one, curious of the reason why a hexagon is better than, say, a cube. Men are capable of demonstration, of justifying their knowledge. In Pappus' words, we are granted a larger portion of σοφία than bees and are more demanding than what an unconscious instinct would allow for: we want something more than what is enough for the bare necessities of life.

The choice of bees is not casual: their political virtues were already known and had been commented upon, as had their geometrical notions.[91] Aelian (second to third century AD), for instance, praises the little insects that are capable of pro-ducing beautiful shapes without τέχνη or rulers or com-passes.[92] Bees were also among the animals used as example in philosophical disputes about the difference between humans and animals, often involving, on opposing sides, Platonists and Stoics. Origen, who tackled the problem in his *Contra Celsum*, is an example quite close in time to Pappus. More-over, bees were a common analogy for purified souls (or were actually seen as such), in Platonist theories which derived from what Plato says about the souls of just people who are

---

[90] Appendix, 1233. Hultsch notes that the passage avoids hiatus.
[91] See the evidence collected in Davies & Kathirithamby (1986).
[92] Aelian, *De natura animalium* V 10–13.

reincarnated, as the best of all possible destinies, into bees and suchlike animals.[93]

Pappus therefore vindicates man's curiosity and capacity for reasoning against the bees' merely instinctive virtue and limited knowledge. The fact that the objects of curiosity are geometrical notions and that the rational reasoning is a capacity for mathematical demonstration allows us to read the contrast between men and bees as a contrast between two kinds of knowledge. One, although true in itself, is not justified by rigorous argumentation, and is therefore opposed to the other (mathematical knowledge), which, by producing such rigorous demonstrations, satisfies man's most distinctive intellectual characteristics.

The argumentative structure of book 5 is thus ingeniously constructed: once convinced by the introduction that it is of man's highest nature to require proofs, the reader will be led into the demonstrations that follow in the text, and will be given several signposts as to what the main results are (they are repeated often enough). He will also be reminded of the author's relation to the famous mathematicians of the past: that Pappus is extremely well acquainted with their results, can quote them, produce variations on them and generally make them easier to understand. A comparison will also be more or less directly invited between philosophers' treatment of the topics and the author's treatment. Megethion ideally should emerge from book 5 having recognized the importance not just of mathematical knowledge, but of its dispenser, producer and expert, i.e. Pappus.

Conclusion

What does the above analysis of book 5 tell us about the wider picture? First of all, that Pappus was not operating in a

---

[93] The evidence for Origen, the Stoics and the Platonists in Chadwick (1947); Plato at *Phaedo* 82a-b. For Platonist theories, e.g. Plotinus, *Enneades* III 4.2.17; Porphyrius, *De antro nympharum* 18; cf. Cook (1895); Willem (1928–30).

vacuum; on the contrary, he was aiming at several targets in the "outside world:" possible or actual pupils, possible or actual competitors, possible third parties who could warrant and uphold his claims to authority and constitute the basis for a widespread reputation. The existence of competition alerts us to the kind of stakes involved: competition, including intellectual competition, implies the need for recognition, the presence of an audience, preoccupation with image, and some interaction, even unwilling interaction, with the outside world.

We can also start to see how Pappus modifies previous materials: apart from some more specifically mathematical features (interest in manipulation of constructions and in subcases), he deploys his sources in a way which is not random or cut-and-paste, but functional to the purpose of his account. Authors are cited, quoted or not mentioned at all, depending on the general effect he wants to achieve. Pappus constructs a past where heroes of mathematics such as Euclid or Archimedes lend prestige and at the same time transmit results; he carefully shapes his own image as the continuator of that past and as the means of access to those authors (as the person through whom those authors can be known); he underlines his own contribution to the tradition, in the form of variations on previous results.

In the next chapter we will look at book 8 of the *Collection*, in order to see how Pappus deals with another of the areas of mathematics we encountered in the first chapter, namely mechanics.

CHAPTER 3

INCLINED PLANES AND ARCHITECTS

Ancient mechanics has been called the "Cinderella of Greek science"[1] – an appropriate definition, if we consider the treatment it has received for years at the hands of historians. Most of us are probably familiar with accounts according to which Greek science, nay, the whole of the ancient world was traversed by a dichotomy between "pure" and "applied"; "theoretical" and "practical."[2] Such a vision relies for support on evidence such as the image of Archimedes conjured up by Plutarch:

To [the engines of war] he had by no means devoted himself as work worthy of his serious effort, but most of them were mere accessories of a geometry practiced for amusement, since in bygone days Hiero the king had eagerly desired and at last persuaded him to turn his art somewhat from abstract notions to material things ... For the art of mechanics . . was first originated by Eudoxus and Archytas, who ... gave to problems incapable of proof by word and diagram a support derived from mechanical illustrations that were patent to the senses.... But Plato was incensed at this, and inveighed against them as corrupters and destroyers of the pure excellence of geometry, which thus turned her back upon the incorporeal things of abstract thought and descended to the things of sense, making use, moreover, of objects which required much mean and manual labour. For this reason mechanics was made entirely distinct from geometry, and being for a long time ignored by philosophers, came to be regarded as one of the military arts."[3]

---

[1] Fraser (1972).
[2] The dichotomy is variously characterized, but inevitably present in Diels (1924); Boyer (1939); Farrington (1946); Koyré (1948a) and (1948b) – it is only fair to say that Koyré himself cites many counter-examples to the evidence commonly deployed; Klemm (1954); Forbes (1960); Finley (1965); Vernant (1965); Krafft (1966); de Santillana (1961); Fraser (1972); Pleket (1973); Gille (1980); Oleson (1984); Parroni (1989). For better balanced views, see the more recent works by Ferrari (1984) and (1985); Franco Repellini (1989); Houston (1989-90); Schürmann (1991); Gara (1994); Traina (1994).
[3] *Vita Marcelli* 14 (305–306). On this see e.g. Authier (1989).

There is quite a lot of other evidence which represents ancient attitudes to practical knowledge in much the same light. Revealing a surprising affinity with some twentieth-century academics, many Greeks described themselves, and the figures they constructed as embodiments of their epistemic ideals (their heroes of knowledge), as fond of philosophy and speculative life, but with little inclination for toil, labour and sweat.

But how representative are these many Greeks, and how reflective of say, the real social position of people involved in activities other than philosophy or speculative life? All too often ancient sources have been uncritically taken as mirrors of an actual state of things, and all too often a selection of the extant evidence has been made to represent *the* ancient view or at least the mainstream or dominant ancient view on the subject of practical knowledge and mechanics. Yet, how does one define the dominant or mainstream view, especially over such a long period of time and across such different realities as those of the western Roman Empire or the Hellenistic Near East? Should our aim as historians be to describe the mainstream view? Does it make sense to speak of a mainstream view at all?

I believe that the future of the history of ancient mechanics will lie less in attempting general descriptions of the Greek and Roman mind than in providing some essential historical groundwork, such as up-to-date assessments of the material evidence, which in many cases has been carefully collected but incorrectly appreciated; studies of key figures such as Hero of Alexandria or Ptolemy; enquiries into the notions themselves of practice, utility and technical knowledge; evaluations of the literary evidence outside the inevitable Plato, Aristotle and Plutarch.

With this chapter I hope to contribute an examination of Pappus' mechanics, as set out in book 8 of the *Collection*. I will present its contents, and relate them to the tradition before him, on which he relies to a great extent, and to the circumstances of the fourth century as I have sketched them in the first chapter. It will be in a sense an interplay of past and present, thus reflecting the interplay in Pappus' text itself. I will show that once again Pappus appropriates results from the past for present purposes, and that the way in which he does it exhibits some typical features, that is, mathematical results are reworked along some characteristic lines. As we did for book 5, we start from the introduction:

The mechanical enquiry, Hermodorus my son, which has many and important uses for things in life, is rightly held by philosophers to be worthy of the highest esteem and is zealously studied by anybody interested in learning, because it is almost the first to deal with the natural enquiry about the matter of the elements in the universe. For theoretical [mechanics] investigates the causes of rest and local motion of bodies, and their movement in space, in their generalities, and not only investigates the causes of whatever moves according to nature, but also moves what goes forcibly against nature from its own place towards a contrary motion; and it contrives to do this by means of enquiries originating from that matter and appropriate to it. The mechanicians around Hero say that there are a discursive and a manual part of mechanics; the theoretical part is composed of geometry, arithmetic, astronomy and discourses about nature, the manual part of work in metals, architecture, carpentry and painting and of manual practice in these. They say that someone who has been trained from a young age in the aforesaid sciences and in addition has reached mastery of the aforesaid arts and has a versatile mind for these things will be the best architect and inventor of mechanical devices. But as it is not possible for the same person to master these studies and to learn the above-mentioned arts at the same time, they advise a person wishing to put his hand to mechanical activities to use what is at hand in his own arts for any necessity.

Of all the arts, the most necessary for the uses of life [mechanics being the leading part of architecture] are: that of the makers of mechanical powers, they themselves being called mechanicians by the ancients (for they lift great weights by mechanical means to a height, contrary to nature, moving them by a lesser force); that of the makers of engines necessary for war, also called mechanicians (for they hurl missiles both of stone and of iron and suchlike objects to a great distance, by means of the instruments, known as catapults, that they make); in addition, the art of those who are in their turn especially called makers of machines (for water is raised from a great depth more easily by means of the instruments for water-drawing which they build). The ancients also call mechanicians the wonder-workers, of whom some practice their art by means of air, as Hero in *Pneumatica*; some by means of strings and ropes, thinking to imitate the movements of living things, as Hero in *Automata* and *Balances*; others by means of bodies floating in the water, as Archimedes in *On Floating Bodies*, or by telling the time by means of water, as Hero in *Hydria*, which seems to have something in common with the enquiry on sun-dials. They also call mechanicians those who know about the making of spheres, who build models of the sky, by means of uniform and circular movement of water. Some indeed say that Archimedes the Syracusan knew the cause and the reason of all these. [The objection by Carpus of Antioch follows.] [G]eometry is in no way injured, but is capable of giving content to many arts by being associated with them, indeed [it, being as it were mother of the arts, is not injured by dealing with construction of instruments and architecture; indeed it is not injured by being associated with land-division, gnomonics, mechanics and scenography], on the contrary, it seems to favour these arts and also to be appropriately honoured and adorned by them.[4]

---

[4] 1022.1–1028.3: Ἡ μηχανικὴ θεωρία, τέκνον Ἑρμόδωρε, πρὸς πολλὰ καὶ μεγάλα τῶν ἐν τῷ βίῳ χρήσιμος ὑπάρχουσα πλείστης εἰκότως ἀποδοχῆς ἠξίωται πρὸς τῶν φιλοσόφων καὶ πᾶσι τοῖς ἀπὸ τῶν μαθημάτων περισπούδαστός ἐστιν, ἐπειδὴ σχεδὸν πρώτη τῆς περὶ τὴν ὕλην τῶν ἐν τῷ κόσμῳ στοιχείων φυσιολογίας ἅπτεται. στάσεως γὰρ καὶ φορᾶς σωμάτων καὶ τῆς κατὰ τόπον κινήσεως ἐν τοῖς ὅλοις θεωρηματικὴ τυγχάνουσα τὰ μὲν κινούμενα κατὰ φύσιν αἰτιολογεῖ, τὰ δ᾽ ἀναγκάζουσα παρὰ φύσιν ἔξω τῶν οἰκείων τόπων εἰς ἐναντίας κινήσεις μεθίστησιν ἐπιμηχανωμένη διὰ τῶν ἐξ αὐτῆς τῆς ὕλης ὑποπιπτόντων αὐτῇ θεωρημάτων. τῆς δὲ μηχανικῆς τὸ μὲν εἶναι λογικὸν τὸ δὲ χειρουργικὸν οἱ περὶ τὸν Ἥρωνα μηχανικοὶ λέγουσιν· καὶ τὸ μὲν λογικὸν συνεστάναι μέρος ἔκ τε γεωμετρίας καὶ ἀριθμητικῆς καὶ ἀστρονομίας καὶ τῶν φυσικῶν λόγων, τὸ δὲ χειρουργικὸν ἔκ τε χαλκευτικῆς καὶ οἰκοδομικῆς καὶ τεκτονικῆς καὶ ζωγραφικῆς καὶ τῆς ἐν τούτοις κατὰ χεῖρα ἀσκήσεως· τὸν μὲν οὖν ἐν ταῖς προειρημέναις ἐπιστήμαις ἐκ παιδὸς γενόμενον κἂν ταῖς προειρημέναις τέχναις ἕξιν εἰληφότα πρὸς δὲ τούτοις φύσιν εὐκίνητον ἔχοντα, κράτιστον ἔσεσθαι μηχανικῶν ἔργων εὑρετήν καὶ ἀρχιτέκτονά φασιν. μὴ δυνατοῦ δ᾽ ὄντος τὸν αὐτὸν μαθημάτων τε τοσούτων περιγενέσθαι καὶ μαθεῖν ἅμα τὰς προειρημένας τέχνας παραγγέλλουσι τῷ τὰ μηχανικὰ ἔργα μεταχειρίζεσθαι βουλομένῳ χρῆσθαι ταῖς οἰκείαις τέχναις ὑποχειρίοις ἐν ταῖς παρ᾽ ἕκαστα χρείαις. Μάλιστα δὲ πάντων ἀναγκαιόταται τέχναι τυγχάνουσιν πρὸς τὴν τοῦ βίου χρείαν [μηχανικὴ προηγουμένη τῆς ἀρχιτεκτονῆς] ἥ τε τῶν μαγγαναρίων, μηχανικῶν καὶ αὐτῶν κατὰ τοὺς ἀρχαίους λεγομένων (μεγάλα γὰρ οὗτοι βάρη διὰ μηχανῶν παρὰ φύσιν εἰς ὕψος ἀνάγουσιν ἐλάττονι δυνάμει κινοῦντες), καὶ ἡ τῶν ὀργανοποιῶν τῶν πρὸς τὸν πόλεμον ἀναγκαίων, κινουμένων δὲ καὶ αὐτῶν μηχανικῶν (βέλη γὰρ καὶ λίθινα καὶ σιδηρᾶ καὶ τὰ παραπλήσια τούτοις ἐξαποστέλλεται εἰς μακρὸν ὁδοῦ μῆκος τοῖς ὑπ᾽ αὐτῶν γινομένοις ὀργάνοις καταπαλτικοῖς), πρὸς δὲ ταύταις ἡ τῶν ἰδίως πάλιν καλουμένων μηχανοποιῶν (ἐκ βάθους γὰρ πολλοῦ ὕδωρ εὐκολώτερον ἀνάγεται διὰ τῶν ἀντλητικῶν ὀργάνων ὧν αὐτοὶ κατασκευάζουσιν). καλοῦσι δὲ μηχανικοὺς οἱ παλαιοὶ καὶ τοὺς θαυμασιουργούς, ὧν οἱ μὲν διὰ πνευμάτων φιλοτεχνοῦσιν, ὡς Ἥρων Πνευματικοῖς, οἱ δὲ διὰ νευρίων καὶ σπάρτων ἐμψύχων κινήσεις δοκοῦσι μιμεῖσθαι, ὡς Ἥρων Αὐτομάτοις καὶ Ζυγίοις, ἄλλοι δὲ διὰ τῶν ἐφ᾽ ὕδατος ὀχουμένων, ὡς Ἀρχιμήδης Ὀχουμένοις, ἢ τῶν δι᾽ ὕδατος ὡρολογίων, ὡς Ἥρων Ὑδρείοις, ἃ δὴ καὶ τῇ γνωμονικῇ θεωρίᾳ κοινωνοῦντα φαίνεται. μηχανικοὺς δὲ καλοῦσιν καὶ τοὺς τὰς σφαιροποιίας [ποιεῖν] ἐπισταμένους, ὑφ᾽ ὧν εἰκὼν τοῦ οὐρανοῦ κατασκευάζεται δι᾽ ὁμαλῆς καὶ ἐγκυκλίου κινήσεως ὕδατος. Πάντων δὲ τούτων τὴν αἰτίαν καὶ τὸν λόγον ἐπεγνωκέναι φασίν τινες τὸν Ἀρχιμήδη ... γεωμετρία γὰρ οὐδὲν βλάπτεται, σωματοποιεῖ πεφυκυῖα πολλὰς τέχνας, διὰ τοῦ συνεῖναι αὐταῖς [μήτηρ οὖν ὥσπερ οὖσα τεχνῶν οὐ βλάπτεται διὰ τοῦ φροντίζειν ὀργανικῆς καὶ ἀρχιτεκτονικῆς· οὐδὲ γὰρ διὰ τὸ συνεῖναι γεωμορίᾳ καὶ γνωμονικῇ καὶ μηχανικῇ καὶ σκηνογραφίᾳ βλάπτεταί τι], τοὐναντίον δὲ προάγουσα μὲν ταύτας φαίνεται, τιμωμένη δὲ καὶ κοσμουμένη δεόντως ὑπ᾽ αὐτῶν, (Trans. in Bulmer-

3.1 War, weights and miracles

Pappus is not a unique case. The authors of our extant texts of mechanics often come up with a definition and a history of their discipline; they seem to be particularly aware that their topic has a past and that it changes over time. Given that attitudes unfavourable to mechanics existed in antiquity (abundantly documented and exemplified by the passage in Plutarch we have quoted), this heightened self-awareness can be seen as a reaction to them, and a reaffirmation of identity in the face of opposition. The contrast, however, should not be cast in clear-cut terms, as upper classes versus lower classes, high culture versus popular culture or written tradition versus orally transmitted forms of knowledge. All these factors are indeed present, but their value as interpretative tools needs to be recognized as problematic. For instance, it is clear from the evidence we have examined that mechanics was associated *both* with τέχνη and the crafts *and* with power and the mightiest human representatives of power.

Belopoietics (the construction of war-engines) is probably the area where the variety of meanings with which the ancients themselves invested mechanics is most apparent. Attitudes to war machines ranged from Demetrius Poliorketes' personal enthusiasm (he was wont to spend his free time building catapults with his own hands) to the Spartan king Archidamus' dismay "when he saw a missile shot by a catapult which had been brought then for the first time from Sicily," and exclaimed, "Great heavens! Man's valour is no more!" There are also Vitruvius' edifying episodes, where human ingenuity prevails over even the most astonishing devices.[5]

Thomas (1967) with my modifications). Hero's text on water-clocks is also mentioned in Pappus' *Commentary on Ptolemy* 89.4–5. Proclus, *In Eucl.* 41.3–42.8 has mechanics subsuming, according to the writings of the ancients, belopoietics, wonder-working, the science of equilibrium in general, sphaeropoietics and astronomy, "which inquires into the cosmic motions, the sizes and shapes of the heavenly bodies" and whose parts are gnomonics, meteorology and dioptrics.

[5] Plutarch, *Vita Demetrii* 20–21; *Regum et imperatorum apophthegmata* 191e and *Apophthegmata Laconica* 219a, respectively; Vitruvius, *De Architectura* 10.16.3–8.

3 INCLINED PLANES AND ARCHITECTS

Writers of belopoietics often explicitly confronted the past, addressing the issue of what had been produced before them. They often indicated that there had been a progress with respect to the past, which could be measured on the basis of performance. Once again, attitudes ranged widely. Procopius observed, with relation to some war engines, "thus, as time goes on, ingenuity is ever wont to keep pace with it by discovering new devices," while according to Frontinus "the invention of [works and engines of war] has long since reached its limit, and for the improvement of [it] I see no further hope in the applied arts." The fifth-century historian and the first-century public officer differ in their opinions, but they both view military mechanics in chronological terms, as something where time, past and present have a role to play.[6]

Philo of Byzantium (early second century BC), who is among Pappus' sources, says that:

the ancients ... did not reach a conclusion, ... since their experience was not based on many works, but they did decide what to look for. The more recent [technicians] drew conclusions from former mistakes, and from things that were experienced after those looking at the constant element, they reached the principle and basis of construction.... Alexandrine craftsmen achieved this first, being heavily subsidized through the interest of kings fond of the arts and of fame (that is, the Ptolemies)[7].

Hero of Alexandria, in his turn, justifies the inclusion in his treatise of outmoded engines on the grounds that it will give his readers a fuller view of the changes undergone by the discipline over time.[8]

In a situation where the past was ever present in some guise or another, certain features became to an extent codified; at

[6] Procopius, *De bellis* 8.11.28; Frontinus, *Stratagemata* 3 Introd.; see also Cracco Ruggini (1985).
[7] *Belopoietics* 50.19–26: τοὺς ... ἀρχαίους μὴ ἐπὶ πέρας ἀγαγεῖν ... οὐκ ἐκ πολλῶν ἔργων τῆς πείρας γεγενημένης, ἀκμὴν δὲ ζητουμένου τοῦ πράγματος· τοὺς δὲ ὕστερον ἐκ τε τῶν πρότερον ἡμαρτημένων θεωροῦντας καὶ ἐκ τῶν μετὰ ταῦτα πειραζομένων ἐπιβλέποντας εἰς ἑστηκὸς στοιχεῖον ἀγαγεῖν τὴν ἀρχὴν καὶ ἐπίστασιν τῆς κατασκευῆς ... τοῦτο δὲ συμβαίνει ποιῆσαι τοὺς ἐν Ἀλεξανδρείᾳ τεχνίτας πρώτους μεγάλην ἐσχηκότας χορηγίαν διὰ τὸ φιλοδόξων καὶ φιλοτέχνων ἐπειλῆφθαι βασιλέων; trans. Marsden (1971), with my modifications.
[8] Hero, *Belop.* 73–74; see Cuomo (forthcoming). Another example where Hero includes "old" results for the sake of completeness is the account of the five mechanical powers in *Mechanica* 2.10.

96

the same time, the existence of common reference points did not mean total uniformity of outlook or general agreement among the sources. Mechanics retained a somewhat "elastic" quality, and the images associated with it varied accordingly: it could be seen as a communal resource of mankind or as the reserve of a few individuals endowed with a special sort of crafty intelligence; it imitated nature but produced things that went against or beyond it; it created machines for war but was also instrumental in promoting peaceful living. In this section I will provide a brief sketch of the textual tradition Pappus is coming from – his past as far as mechanics is concerned.

Our earliest extant account is the pseudo-Aristotelian *Mechanical Questions* (third century BC), which is ascribed to some Peripatetic scholar, maybe Strato of Lampsacus. The treatise examines a number of amazing phenomena, both natural and artificial, and attributes the cause of them all to the extraordinary properties of the circle. From a methodological point of view, the introduction says, mechanical problems are to be placed halfway between physics and mathematics, as they neither coincide with, nor are completely alien from, physical problems. Their "how" (presumably the discourse of mechanics) is made clear by means of mathematics, but the "about what" (the object of mechanics as something which is out there in the world) is made clear by means of physics (that is, by means of natural inquiries).[9]

---

[9] [Aristotle], *Mech.* 847a25–30: Θαυμάζεται τῶν μὲν κατὰ φύσιν συμβαινόντων, ὅσων ἀγνοεῖται τὸ αἴτιον, τῶν δὲ παρὰ φύσιν, ὅσα γίνεται διὰ τέχνην πρὸς τὸ συμφέρον τοῖς ἀνθρώποις.... ὅταν οὖν δέῃ τι παρὰ φύσιν πρᾶξαι, διὰ τὸ χαλεπὸν ἀπορίαν παρέχει καὶ δεῖται τέχνης. διὸ καὶ καλοῦμεν τῆς τέχνης τὸ πρὸς τὰς τοιαύτας ἀπορίας βοηθοῦν μέρος μηχανήν.... ἔστι δὲ ταῦτα τοῖς φυσικοῖς προβλήμασιν οὔτε ταὐτὰ πάμπαν οὔτε κεχωρισμένα λίαν, ἀλλὰ κοινὰ τῶν τε μαθηματικῶν Θεωρημάτων καὶ τῶν φυσικῶν· τὸ μὲν γὰρ ὡς διὰ τῶν μαθηματικῶν δῆλον, τὸ δὲ περὶ ὃ διὰ τῶν φυσικῶν (Our wonder was excited, firstly, by phenomena which occur in accordance with nature but of which we do not know the cause, and secondly by those which are produced by art against nature for the benefit of mankind.... When, therefore, we have to do something contrary to nature, the difficulty of it causes us perplexity and art has to be called to our aid. The kind of art which helps us in such perplexities we call mechanics.... [Mechanical problems] are not quite identical nor yet entirely unconnected with natural problems. They have something in common both with mathematical and with natural speculations; for while mathematics demonstrates the how, natural science demonstrates the about what), Loeb trans. with my modifications. I thank R. Netz for helping me with the sense of the last sentence.

Mechanics itself identifies a set of questions encountered in the course of human experience, or devices produced by human τέχνη in order to obtain some advantageous effect. Of such nature are all the cases where the lesser prevails over the greater, as when a small force moves a big weight.

While remaining close to the original meaning of μηχανή (trick, device), the pseudo-Aristotelian notion of mechanics emphasizes its peculiar capacity to generate surprise and wonder. This is not restricted to phenomena against or beyond nature (παρὰ φύσιν); things that happen according to nature can create amazement too, insofar as their cause is hidden, so they can be ascribed to the realm of the mechanical as well. For instance, the fact that pebbles on the sea-shore are round even though they were originally long stones or shells, or the fact that when people stand up they must make an acute angle between their lower leg and thigh, otherwise they cannot rise at all.[10] The second distinctive feature of mechanics is its usefulness for humankind, the fact that it is, largely speaking, on man's side against an often hostile environment, where humans have to resort to their cunning (think of the πολυμήχανος Odysseus)[11] in order to meet their needs. The *Questions* does not propose new applications of the simple machines it describes, nor does it devise improved engines; it examines natural phenomena or well-known aspects of everyday technology. Indeed, nothing in it hints at an audience whose interest in mechanics was any different in kind from an interest in biology or meteorology, or any other aspect of reality, paradoxical as it might be. Mechanics was in a sense already there – it just needed to be made into an object of knowledge.

Although not far removed in time, the mechanics we find in Philo of Byzantium comes across as rather different, thus reflecting different circumstances of production and probably a different audience. The author was active as an engineer in

---

[10] [Aristotle], *Mech.* 852b23–28 and 857b21-858a2, respectively. See also Ferrari (1985) and Franco Repellini (1989).

[11] Homer, *Iliad* 2.173; 4.358; 8.93. Cf. Detienne & Vernant (1974), 18, 32.

Rhodes and Alexandria, and produced what has been seen as the first attempt to systematize mechanics as a form of knowledge. The *Mechanical Syntaxis*, now mostly lost, originally included chapters on pneumatics, belopoietics, harbour-building, self-moving devices (αὐτόματα) and "stratagems", which perhaps here for the first time were brought together as branches of the same discipline.[12] In the book on belopoietics (which, together with the one on pneumatics and a few other fragments, is all that remains by him), Philo seems genuinely concerned that his recipient, Ariston, should be able, following his indications, actually to build, reduce or enlarge a variety of military machines. Remarks about the strong or weak points of a certain engine abound, as do accurate reports on measurements, dimensions, and choosing the best materials to use. The feeling that we are dealing with real machine-construction is reinforced by Philo's concern for the financial side of things – he observes that the generous patronage given to engineers by the Alexandrine kings was a determining factor in their remarkable progress in the field.[13]

But Philo is also particularly attentive to what we could call the epistemic status of the discipline, and investigates it from a historical point of view. Handed-down knowledge is fundamental: for instance, he can trust himself to recommend a certain adjustment because he has seen a famous engineer in Rhodes employ it.[14] Also, in the passage quoted above, he describes how engineers came in time to the fundamental discovery that the diameter of the hole that holds the spring of the catapult is its "module unit", the piece whose dimensions one needs to refer to if one wants to modify or reproduce catapults keeping their proportions. This major piece of knowledge was the combined result of favourable material conditions (the support of the Ptolemies), theoretical knowledge, such as being able to find two mean proportionals, and opportunely conducted trial-and-error experiences.[15]

---

[12] Ferrari (1984).   [13] Philo, *Belop.* 50.24–26.
[14] *op. cit.* 51.10–14.   [15] *op. cit.* 50.26–29.

Philo's weapons exhibit the characteristics typical of mechanics as described by the pseudo-Aristotle passage: utility and wonder-working, the capacity to produce both amazing and useful effects. Philo, in fact, takes care to preserve a wondrous element in his catapults – they have to look right for their purpose, as well as work properly, and the awesome appearance of the finished product is an integral part of the project. Thus, for instance, Philo recommends that the front of a catapult be equipped with a metal cover to make it look "not unimposing" (ἥσσων).[16] It is made rather clear that one of the functions of belopoietics – that is, of mechanics at the service of war – is to project to the enemy an image of superhuman might, of amazing power achieved without manifest effort.

We find many episodes in ancient history where the spectacular quality of mechanics is associated with, and nearly coincides with, the exercise of power. Think of the effect achieved by Archimedes on the Romans not long before Philo's time (*ca.* 212 BC), again as described by Plutarch:

[The Roman commander Marcellus declares] "Let us stop … fighting against this geometrical Briareus, who … with the many missiles which he shoots against us all at once outdoes the hundred-handed monsters of mythology." … At last the Romans became so fearful that, whenever they saw a bit of rope or a stick of timber projecting a little over the wall, "There it is," they cried, "Archimedes is setting some engine upon us," and turned their backs and fled.[17]

Sometimes the connection between mechanics and power is in the explicit form of political rulers featuring as commissioners, onlookers or direct protagonists of mechanical feats. Among the latter, the Hellenistic king Demetrius Poliorketes (336–283 BC) owed his name (the Besieger) at least partly to the edge given him in sieges by some spectacular machines, above all the *helepolis*, a siege tower so wondrous that even enemies were struck by its beauty and asked Demetrius for a truce specifically in order to have a look at it. Demetrius'

---

[16] *op. cit.* 61.29–62 ff.; 66.17 ff.
[17] Plutarch, *Vita Marcelli* 17 (307).

fame stretched as far as Pappus' time: he is mentioned by
Ammianus Marcellinus, in his own description of war engines
(which is part of the narration of the emperor Julian's cam-
paigns).[18] Dionysius of Syracuse is even credited with the invention of the catapult, at a definite time in history (around
399 BC).

After collecting many skilled workmen, he divided them into groups in accordance with their skills, and appointed over them the most conspicuous
citizens ... the making of great quantities of arms went on apart from ...
public places, in the most distinguished homes. In fact the catapult was
invented at this time in Syracuse, since the ablest skilled workmen had
been gathered from everywhere into one place. The high wages, as well
as the numerous prizes offered to the workmen who were judged to be the
best, stimulated their zeal. And over and above these factors, Dionysius
circulated daily among the workers, conversed with them in kindly fashion, and rewarded the most zealous with gifts and invited them to his
table.[19]

The most famous onlooker of mechanical feats has to be
that other famous tyrant of Syracuse, Hiero. In a (probably
fabricated, but nonetheless poignant) anecdote, he is described
as overseeing the launch of a gigantic ship, which Archimedes
managed to send into the sea all by himself, with a little help
from a mechanical device of the type we will see later described by Pappus. The story acquires an even more interesting dimension if we consider that the ship was a present from
king Hiero to one of the Ptolemies of Egypt, and that the
whole thing was organized as a sort of public celebration
for the Syracusan populace, with Hiero himself presiding over
the show. Of course, the more astounding the present and
the circumstances surrounding it, the more prestige to the
giver.[20]

Archimedes was credited with another mechanical wonder:

---

[18] See especially Plutarch, *Vita Demetrii* 20–21; see also descriptions of the *helepolis*
in Biton, *Belopoietica* 52 ff. (third quarter of third century BC) and Diodorus,
Bibliotheca 20.48; 20.91 (chapter 20 is mostly about Demetrius). Ammianus
Marcellinus at *Res gestae* 23.4. On these issues cf. Gabba (1984); Schürmann
(1991).
[19] Diodorus, *Bibliotheca* 14.41–42.
[20] Plutarch, *Vita Marcelli* 14.8; Athenaeus, *Deipn.* 5.207b. On Hiero see Berve
(1959).

a sphere which reproduced the motions of the heavens. The construction of spheres was one of the disciplines traditionally considered part of mechanics, and it is included in Pappus' definition. Indeed, Pappus reports the attribution to Archimedes of a whole treatise on *Sphere-making*.[21] That the story of the sphere lived on well into the fourth century is testified also by Claudian (AD 370–410), who wrote a short poem to the effect that "[w]hen Jove looked down and saw the heavens figured in a sphere of glass he laughed and said to the other gods: 'Has the power of mortal effort gone so far? Is my handiwork now mimicked in a fragile globe? An old man of Syracuse has imitated on earth the laws of the heavens, the order of nature, and the ordinances of the gods.'" The building of an orrery was the ultimate prodigy: both κατὰ φύσιν, in that it did nothing more than copy nature, and παρὰ φύσιν, in that it brought out the Promethean implications of mechanics, its overtone of hybris.[22]

To move on to Pappus' next and principal source, Hero of Alexandria's version of mechanics is particularly distinctive in that he makes strong statements about its political and philosophical significance. The introduction to the *Belopoietics* is a key text in this respect: it positively claims a primary role for mechanics in the acquisition of the ideal state of ἀταραξία (tranquillity, imperturbability) which was advocated by many of the philosophies of the time. Hero is adamant in denying that mere discourse will ever give anyone real tranquillity. In a world where enemies lurk both within and without the community, words will never assure safety, which instead can granted by mechanics. If, thanks to belopoietics, a city is well equipped against attacks, the peace of mind of its citizens will be securely founded. Hero envisages a kind of psychological warfare, where the enemies will give up their plots not in front

[21] 1026.9–12.

[22] The sphere first mentioned in Cicero, *Tusculanae Disputationes* I XXV 63 (where he compares Archimedes to the Platonic demiurge); *De Re Publica* I 21 XIV; *De Natura Deorum* II 34.88–35.89. Cf. also, among others, Lactantius, *Divinae Institutiones* II 5.18; Firmicus Maternus, *Math.* 1.5 and II 30.26 and Claudian, *Carmina Minora* 51.1–6. For the overtones of hybris, see Vernant (1965); for the topic in general see Aujac (1970); de Solla Price (1974).

of an *actual* use of the weapons to curb them, but because they see the weapons and know that they could *potentially* be used. In other words, the image of power functions as real power.[23]

Hero's awareness of the spectacular potentialities of mechanics is manifested in the *Pneumatics* and the *Automata*, where he describes devices by means of which one can make the gates of a temple open or the fire on an altar set alight "by themselves." These feats of effortless power were meant to be carried out in various public contexts, for instance the religious environment of a temple, in order to impress the masses, or upper class entertainments like a banquet, in order to impress the guests. The wondrous character of mechanics can be traced even in less obvious contexts. In the *Mechanics*, for instance, Hero states that

we will wonder at things which, when we have proved them, are the contrary of what is manifest to us. But the things, the causes of which we can talk about only according to the simplest principles, will increase our wonder even more when we see that the things which we employ are the contrary of what we are used to and what we hold for certain.[24]

---

[23] Hero, *Belop.* 71 ff: Τῆς ἐν φιλοσοφία διατριβῆς τὸ μέγιστον καὶ ἀναγκαιότατον μέρος ὑπάρχει τὸ περὶ ἀταραξίας, περὶ ἧς πλεῖσταί τε ὑπῆρξαν ζητήσεις παρὰ τοῖς μεταχειριζομένοις τὴν σοφίαν καὶ μέχρι νῦν ὑπάρχουσι· καὶ νομίζω μηδὲ τέλος ποτὲ ἕξειν διὰ τῶν λόγων τὴν περὶ αὐτῆς ζήτησιν. μηχανικὴ δὲ ὑπερβᾶσα τὴν διὰ τῶν λόγων περὶ ταύτης διδασκαλίαν ἐδίδαξε πάντας ἀνθρώπους ἀτεράχως ζῆν ἐπίστασθαι δι’ ἑνὸς καὶ ἐλαχίστου μέρους αὐτῆς, λέγω δὴ τοῦ κατὰ τὴν καλουμένην βελοποιίαν, δι’ ἧς οὔτε ἐν εἰρηνικῇ καταστάσει ταραχθήσονταί ποτε ἐχθρῶν καὶ πολεμίων ἐπανόδοις, οὔτε ἐνστάντος πολέμου ταραχθήσονταί ποτε τῇ παραδιδομένῃ ὑπ’ αὐτῆς διὰ τῶν ὀργάνων φιλοσοφίᾳ ... καὶ οἱ ἐπιθυμοῦντες ἐπιβουλεύειν, ὁρῶντες τὴν περὶ αὐτὰ γιγνομένην αὐτῶν διατριβήν, οὐκ ἐπελεύσονται. (The largest and most essential part of philosophical study is the one about tranquillity, about which many researches have been made and still are being made by those who pursue learning; and I think research about tranquillity will never reach an end through reasonings. But mechanics has surpassed teaching through reasonings on this score and taught all human beings to live a tranquil life by means of one of its branches, and the smallest – I mean, of course, the one concerning the so-called construction of artillery. By means of it, when in a state of peace, they will never be troubled by reason of resurgences of adversaries and enemies, nor, when war is upon them, will they ever be troubled by reason of the philosophy which it provides through its engines ... those who wish to plot, observing how they became engaged in the study of the subject, will not attack), trans. in Marsden (1971), with my modifications. I thank D. Sedley for helping me with the sense of this passage.

[24] Hero, *Mech.* 1.7 and 2.33, respectively.

One example is probably the wondrous cog-wheel machine which will enable a single man, "or even a boy," to lift a weight of a thousand talents with a force equivalent to only five talents – the same machine will be described by Pappus.[25]

3.2 Bodies, motion and human limits

Pappus presents mechanics right from the beginning as a θεωρία on which universal approval and praise are bestowed. Philosophers cannot have enough of exalting it; all the mathematicians without exception have devoted their best energies to it; it is useful for many great things in life (a quality for which Archimedes' *Measurement of the Circle* was also praised), and it deals with a vast array of subjects.[26] According to Pappus' first, and probably foremost, definition, mechanics is in particular an inquiry into the movement, rest and local motion of bodies in the universe. It studies the causes of phenomena that happen according to nature (κατὰ φύσιν) and also determines the occurrence of phenomena that go against nature (παρὰ φύσιν), i.e. movements in a direction away from the body's natural site. It thus has both a productive part – it acts on nature – and a cognitive one – it aims to understand nature.

The general framework of Pappus' definition could be termed, largely speaking, Aristotelian: this impression is given by the occurrence of notions such as the search for causes, the natural sites of bodies and the two categories of κατὰ and παρὰ φύσιν. We also find a similar characterization of mechanics in Vitruvius, who links it with cosmology. Nevertheless, it is unlikely that Pappus drew on Vitruvius directly, as Vitruvius seems not to have been known in the late ancient Near East; as for Aristotle, some of his ideas or buzz-words had by then entered generic discourses about nature and the

---

[25] Hero, *Mech.* 1.1.
[26] 1022.1–8; the reference to Archimedes in Eutocius, *In Archimedis Dimensionem Circuli* 228.20–1.

kosmos to an extent too great for the echoes in Pappus to be significant of a direct link.[27]

As a possible source for the definition in book ℰ I would be inclined to suggest some Ptolemaic work. Ptolemy is actually quoted by Pappus, who says that his *Mathematics* contains all the basic concepts about bodies and motion that the reader needs, about "what is the heavy and what is the light, and what is the cause of the movements of the bodies upwards and downwards, and of the upwards and downwards themselves, what idea one has of them and by what limits they are defined."[28] A definite source is Hero, on whose authority Pappus distinguishes two parts to mechanics. The first is discursive or argumentative (λογικόν), and includes several sciences (ἐπιστῆμαι), the second is accomplished through arts (τέχναι) and is practical, or rather manual (χειρουργικόν). The combination of ἐπιστήμη and τέχνη reappears later, when mechanics is referred to as "consisting of science and art together."[29] I would like to stress that Pappus makes no difference of status between the two parts. He must have been aware, however, that their relation to each other was an issue for debate,

---

[27] For Aristotle, see e.g. *De Caelo* 268b14 ff.; Vitruvius, *De Architectura* 10.1.4: "All machinery is generated by nature, having as guide and teacher the revolution of the universe. For first indeed, unless we could observe and contemplate the continuous motion of the sun, moon and also the five planets, ... we should not have known their light in due season ... Since then our fathers had observed this to be so, they took precedents from Nature; imitating them ... they developed the comforts of life by their inventions." The similarities between Pappus and Vitruvius might be explained by common sources, maybe Posidonius, or by Pappus relying on Hero, who may have known Vitruvius (see Fleury (1994)), and mentions Posidonius, *Mech.* 1.24.

[28] 1030.1–4: Τί μὲν οὖν ἐστιν τὸ βαρὺ καὶ τὸ κοῦφον, καὶ τίς αἰτία τῆς ἄνω καὶ κάτω τοῖς σώμασι φορᾶς, καὶ αὐτό γε τὸ ἄνω καὶ κάτω τίνος ἐννοίας ἔχεται καὶ τίσιν ἀφώρισται πέρασιν. The *Mathematics* probably corresponded to the *Almagest* or to some version of the *Almagest*. For these themes in Ptolemy, see *Syntaxis* I 7; *Planetary Hypotheses* 2.1 ff.; see also Taub (1993). Micheli (1995), 132, thinks that Pappus' mechanics is inspired by Aristotle except for the cosmological undertones.

[29] 1028.4–5: τῆς μηχανικῆς ἐπιστήμης ὁμοῦ καὶ τέχνης ὑπαρχούσης Jerome (Hieronymus), *Epistulae* 53.6 (AD 394–396), has an analogous distinction between grammarians, rhetoricians, philosophers, geometers, doctors and astrologers, whose knowledge is divided into three parts (doctrine, method and experience), and practitioners of "minor arts which are practiced not so much with discourse as with the hand" ("artes minores quae administrantur non tam λόγῳ quam manu"), such as smiths, farmers and carpenters.

because quickly, as if to prevent possible criticisms, he points out some complications. Mechanics in its entirety, he says, is too vast for a single individual to master both its aspects. Not that the desirable thing is to avoid the manual part altogether, or to delegate it to the lower artisans – on the contrary, Pappus stresses that the best thing would be to combine *both* practical *and* theoretical knowledge, but that this is very, very difficult to achieve in one lifetime. Thorough competence in all the aspects of mechanics is thus elevated to the level of an ideal, embodied by the great Archimedes, who, "as they say," managed to know the cause and reason of all its different branches. The indirect mode of the testimony adds to the legendary tone – perfection belongs to the realm of the heroes of knowledge, and ordinary people can surely be excused if they do not attain it.[30]

Pappus himself was no engineer – nowhere does he claim to have first-hand experience of, or to have personally built, the (only) device, a weight-lifting machine, that he describes. There must be more to his defence of the complementarity of the two aspects of mechanics than a reference to his own activities. In my view, Pappus' position is to be read, at least partly, as a matter of *esprit de corps*, a statement of allegiance to the mathematicians who had engaged in mechanics before him, such as Archimedes himself, whose depictions in the fourth century include Claudian's poem (quoted before) and Firmicus Maternus' horoscope, where he figures as the quintessential "divine inventor" of mechanical devices, including the famous sphere;[31] or, for instance, the various ancients whose definitions Pappus reports. Some defined mechanics as the art of lifting weights against nature (παρὰ φύσιν);[32] some

[30] Many of these themes, including having Archimedes as the ideal embodiment of all-round knowledge, practical *and* theoretical, are found in Vitruvius' *De Architectura*; cf. Goguey (1978); Romano (1987).
[31] Firmicus Maternus, *Math.* II.148.11–27. The twelfth-century writer Johannes Tzetzes introduces Philo of Byzantium, Archimedes, Hero of Alexandria and Pappus as mechanicians, *In Aristophanis Nubes* 621–622 and *Chiliades* 2.35.
[32] The so-called μαγγαναρίος τέχνη. One of the meanings of μάγγανον is "magic" (*LS* under μάγγανον); cf. Eunapius, *Vitae* 474, for use of the term with this meaning. The word also occurs in Hero to indicate a part of the pulley, *Belop.* 84.12 (this reference in Bulmer-Thomas (1967), 616n).

thought that belopoietics could unqualifiedly be called mechanics. Some, on the other hand, thought that proper mechanics concerns itself with water-pumps. The ancients also identified as mechanicians the wonder-workers, including people engaged in sphaeropoietics, hydrostatics and clock-making. Here Pappus finds it necessary to specify that the wonder-workers (θαυμασιουργοί) only *"seem to imitate"* the movements of animated beings (in case anyone had been taken in by their tricks) and that their puppet theatres work with the help of ropes or strings. The magic is thus dispelled by lifting the veil to reveal the purely physical phenomena behind it.

In fact, Pappus rehearses his own definition of mechanics (an enquiry related to physical causes) for most of the previously established identities of the discipline. For each of them, he provides an explanation in terms of the specific physical problem involved. Thus, belopoietics is characterized as the problem of how to throw projectiles of various kinds a long distance away, while sphaeropoietics, with its quasi-mystical "demiurgical" connotations, is presented as simply a matter of regular movements of a water-operated mechanism, analogous to those of water-clocks. Pappus' version is more general and subsumes previous definitions, which all focused on what he regards as mere sub-branches.

We can now try to sum up the features of Pappus' presentation of mechanics: he first puts forth a definition of his subject that fits the mould of the tradition, in that it places mechanics between mathematical and physical theory, in a domain which includes phenomena both in accordance with nature and against nature. He suggests continuities with Ptolemy, (whom I take to have been a well-respected author at the time) both implicitly, with his emphasis on the cosmological aspects of mechanics, and explicitly, with his mention of Ptolemy's book. The weak points are underplayed: the heterogeneity of the discipline is strategically recognized, but then quickly shifted on to the level of pedagogical ideal. The fact that the discursive and manual parts are not easily found together is not because the former is superior, but because education and training have unavoidable limits: there is a limit to how much an individual can actually learn. The various definitions of mechanics handed down from the tradition are collected under the general umbrella of Pappus' physical enquiry, and the connotations of wonder-working are downplayed. Finally, by evoking the never-to-be equalled figure of Archimedes, Pappus associates the idea of comprehensive mechanics with the myth of the perfect scientist, and cashes out on contemporary perceptions of that myth.

Pappus' emphatic praise makes one wonder whether he is not overdoing it in order to counterbalance possible detractors; in other words, whether his laudatory tone is trying to cover voices to the contrary. In fact, one voice to the contrary is given expression: Carpus of Antioch, who was concerned about the interaction of mechanics and geometry and claimed that Archimedes did not deem mechanics worth writing about.[33] The tone of Pappus' response is more patronizing than polemic: he registers Carpus' assertion and observes that, nevertheless (καίτοι, a concessive clause), Archimedes was considered superhuman and was universally praised precisely because of his achievements in mechanics. As for his writings, he was extremely concise and painstaking even when composing treatises on geometry and arithmetic. (The implications of this are not drawn out; maybe Pappus wants to say that Archimedes was rather fussy about what he published anyway.) On the other hand (comes the riposte) do not Carpus himself "and some others" combine geometry and arts to good effect? Indeed, geometry and mechanics benefit each other greatly, as Pappus articulates at some length.[34] This latter passage, which concludes the rebuff of Carpus, has been deemed spurious (unnecessarily, I think) by Hultsch, probably because of its vaguely mystical tones. Geometry is described as the "mother" of agrimensure, scenography and mechanics, who dutifully honour and beautify (literally "cosmify") her –

---

[33] Possibly the same Carpus "the engineer" mentioned by Proclus, *In Eucl.* 125.25 and 241.19.
[34] 1026.6–1028.3.

one is reminded in fact of some late ancient representations of the mathematical disciplines in terms of personified semi-divinities.[35]

3.3 Wheels, triangles and columns

The introduction of book 8 draws up an entire programme for mechanics; possible objections are dismissed on the strength of well-orchestrated rhetorical points, or by strategically using the aura of big names like Archimedes. The cosmological characterization manages to give some appearance of unity and order to the various branches of mechanics, relating them to the irreproachable pursuit of *rerum cognoscere causas*, on the basis of which all the previous angles on the discipline should be reconsidered in terms of physical problems.

Pappus proceeds with his systematization by arranging the contents of his account under three general headings, all set in the form of problems: the inclined plane (given a weight moved by a given force in a horizontal plane, find the force necessary to move it up a plane inclined to the horizon at a given angle); the duplication of the cube (find two mean proportionals to two given lines); and the cog-wheel (find whether it is possible, given a cog-wheel with a given number of teeth, to apply to it another cog-wheel with a given number of teeth – also find the diameter of this latter).[36] The problems, as we will notice in the course of the discussion, had been previously discussed by Hero in his *Mechanics* (extant in a ninth-century Arabic translation).[37]

"Each of these [problems] in its own way will manifest itself to be useful, together with the other [problems], to the architect and the mechanician." Indeed, Pappus takes care to append to two of them a clause clarifying its overall significance:

---

[35] See e.g. Martianus Capella's *De nuptiis Philologiae et Mercurii* (last quarter of the fifth century AD).

[36] 1028.4–27.

[37] Precedents for the inclined plane in Hero, *Mech.* 1.20 ff., especially 23; the two mean proportionals at *ibid.* 1.11; the cog-wheel mechanism (but not the problem about the quantity of teeth) at *ibid.* 1.1.

the problem of the inclined plane is presented as useful to the mechanicians engaged in the μαγγαναρίος τέχνη; the problem about the cog-wheel (τύμπανος, i.e. shaped like a drum) is labelled, again, useful for many things in the art of instrument-building.[38] On the whole, the three problems must have covered quite a significant portion of the traditional domain of mechanics, especially since Pappus has referred the reader to other works where subjects like wonder-working are concerned. In this light, he feels entitled to claim that his own account, as well as being more concise and clearer, is altogether "better" than the previous ones.[39]

Pappus' "program" is actually carried out a bit at random – the statement of the three problems is a sort of false start, after which he lists a number of things that he does not need to talk about, because, he says, they can already be found in Ptolemy; he provides a definition of centre of gravity, in order to elucidate other theorems that deal with the subject of κεντροβαρικὴ πραγματεία (he refers to this as a summary) and sets out several propositions which he has decided to report even though they are already contained in Archimedes and Hero. The addressee himself, Hermodorus, is invited to learn from their books, respectively the *Equilibria* and the *Mechanics*, but Pappus does not expect many people to be familiar with them.[40]

Thus, while declaring his dependence on previous texts,

---

[38] The passage quoted at 1028.27–29: ἕκαστον δὲ τούτων ἐν τῷ οἰκείῳ τόπῳ γενήσεται φανερὸν μετὰ καὶ ἄλλων χρησίμων ἀρχιτέκτονι καὶ μηχανικῷ. The clarifying clauses at 1028.15–18 and 25–27, respectively. Cf. Downey (1947–8), who thinks that with the introduction to book 8 Pappus intended to provide a curriculum for architectural studies.

[39] 1028.8–10: "and I shall write down the theorems conveniently found by us in a manner more concise and clearer and put them together in a better order than the one in which others have written them down before" (τὰ ὑφ' ἡμῶν εὐχρήστως ἀνευρημένα θεωρήματα συντομώτερον καὶ σαφέστερον ἀναγράψαι βελτίονί τε λόγῳ τοῦ παρὰ τοῖς πρότερον ἀναγεγραμμένου συντάξαι). The claim is, of course, a common one in "technical" and non-technical treatises alike: see e.g. Hero, *Mech.* 3.1.

[40] 1030.6–10 for the introduction to the "summary" (συνέχον); the introduction to the other propositions is at 1030.18–1034.6. The reference, at least for the modern reader, is to Archimedes, *Planorum Aequilibria* I 9–10; 13–14 and Hero, *Mech.* 2.35–36; 38–40.

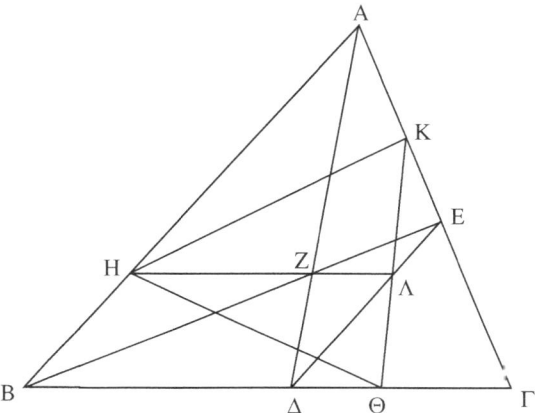

Fig. 3.1. In the triangle ABΓ, we posit H, Θ and K such that AH:HB = BΘ:ΘΓ = ΓK:KA. The points Δ and E are the mid-points of BΓ and ΓA. The point Z, which is the intersection point of AΔ and BE, will be the centre of gravity of ABΓ. After a long series of demonstrative steps involving parallels, proportions and the two lemmas we mentioned, it is proved that Z is the centre of gravity of HKΘ as well.

Pappus emphasizes his own explanatory or mediating function. His propositions on the centre of gravity of a triangle are an interesting example of the way in which he deals with his sources, so we will look at them in some detail. The first theorem aims to prove that the triangle ABΓ and the triangle HΘK internal to it, as shown in fig. 3.1, have the same centre of gravity. The proof appeals to two auxiliary lemmas, whose demonstration is postponed, a thing not unusual for Pappus, as we have seen with book 5.[41]

The first part of the proposition is a demonstration that Z is the centre of gravity of the triangle ABΓ. This is probably drawn from Hero's *Mechanics* – the two argumentations basically coincide.[42] A short passage ("One has to imagine that the point Z, as we said, is in the middle of the triangle ABΓ,

---

[41] 1034.7–1038.4. The two lemmas at 1038.5–1040.10 and 1040.11–25.

[42] Hero, *Mech.* 2.35. Hero's editor, W. Schmidt, endorsed a Heronian provenance for Pappus' result to the point of including the text from the *Collection*, with some modifications, in his edition of the Greek fragments of the *Mechanics*, 292.1–294.3.

which is clearly assumed to be of homogeneous thickness and weight") is however expunged by Hultsch. The specification that the triangle is to be of homogeneous thickness and weight is in Hero's text – rather than an interpolation, then, it would seem reasonable to think that Pappus here is simply drawing on his source, and underlining the significance of the study of centres of gravity for architects and mechanicians, in that the geometrical objects in question are presented as related to objects in the material world.[43]

Now, while it is quite indisputable that the last part of the *Collection*[44] is taken *verbatim* from Hero, I think that a different kind of operation is under way in this part of book 8. Hero's demonstration of the centre of gravity of a triangle is used for the first part of Pappus' proposition. The two auxiliary lemmas and the additional statement that the two triangles have the same centre of gravity, on the other hand, are not contained in Hero and fit with what I have indicated as features characteristic of Pappus' practice: they focus on how the construction of a problem relates to the data given at the outset and how it changes when the given data change.

At the end of the passage on centres of gravity Pappus specifies: "these and similar things on the one hand contain [the] theory, while things with the capacity to have mechanical utility would be of such and such a kind."[45] This echoes an earlier passage where Pappus has emphasized that, mechanics being composed both of sciences and arts, he was going to supply some geometrical investigations handed down by the ancients and by him judged useful for the task at hand.[46] I take these as statements of the view that mechanics and mathematics are complementary: mechanics looks at some objects in the world as objects of knowledge – knowledge which is mathematically founded – and builds useful devices

---

[43] 1034.22–24: νοεῖν δὲ δεῖ τὸ Ζ, ὡς προείρηται, κείμενον ἐν μέσῳ τοῦ ΑΒΓ τριγώνου ἰσοπαχοῦς τε καὶ ἰσοβαροῦς δηλονότι ὑποκειμένου. Hero, *Mech.* 2.35: "How one finds the centre of gravity of a homogeneously thick and heavy triangle."

[44] 1114.22 ff.

[45] 1046.26–7: Ταῦτα μὲν οὖν καὶ τὰ τοιαῦτα θεωρίαν ἔχει, τὰ δὲ καὶ εἰς χρείαν δυνάμενα πεσεῖν μηχανικὴν τοιαῦτ' ἂν εἴη.

[46] 1028.4–9.

on the basis of such knowledge. In other words, the enquiry about centres of gravity concerns *both* theory, because it belongs to the basic principles of mechanics as codified by Archimedes, Hero and Ptolemy, *and* practice, because it is on those bases that one can then, for instance, lift weights up an inclined plane.

In fact, the problem of the weight moved up an inclined plane comes next, preceded by a proposition on how to determine the angle at which a plane is inclined to the horizon. The discussion is divided into two sub-cases (another feature which we have already indicated as characteristic of Pappus).

After a general geometrical demonstration, the problem is repeated in the form of what Pappus calls an example (παράδειγμα) of both construction and demonstration: they are repeated, but this time specific numbers are given for the weight to be moved, the force available (expressed in number of men) and the inclination of the plane.[47] We will look at this example in more detail in chapter five.

Pappus' solution to the problem of the inclined plane is wrong on modern criteria, because it assumes that one needs a given force to move a body along a horizontal plane, while, according to modern physics (in an ideal situation where the body moves in a vacuum and there is no friction), it takes an infinitesimally small force to move such a body.[48] Hero himself had stated in the *Mechanics*: "Some people think that a weight lying on the ground will be moved only by a force equivalent to it, because they trust false appearances. We ... prove that a weight situated in the abovesaid position will be moved by a force smaller than any known force." In the discussion that follows, Hero underlines factors such as smoothness of all the surfaces involved, cohesiveness of material

---

[47] The general demonstration at 1054.4–1056.29; the example at 1056.30–1058.26.

[48] Cf. Ver Eecke (1933b); Drabkin & Drake (1969). Pappus' solution was adopted by some sixteenth-century mathematicians, e.g. Guidobaldo dal Monte, in preference to a more correct one, which was also available to them, apparently because the latter had the disadvantage of coming from writers less prestigious than the classic author Pappus. The choice of Pappus' construction would then be explained in terms of authority and status rather than of its internal truth or error; cf. Biagioli (1989). Note that Hero's *Mechanics* was first published in 1893.

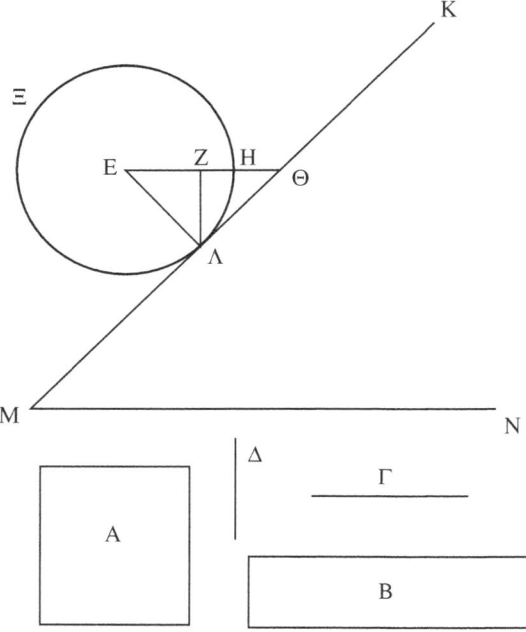

Fig. 3.2. The sphere with centre E is taken to have the same weight as a weight A which needs the force Γ to be moved along a horizontal plane. The triangle EΛZ is given "in species," i.e. its angles are given, and the ratio ZH:EZ is also given. We posit that this ratio is equal to the ratio of the weight A to another weight B, and also equal to the ratio of the force Γ to another force Δ. But the force Γ could move the weight A along a horizontal plane, therefore the force Δ will be able to move the weight B along the same plane. If then we put the centres of gravity of the two weights at the points E and H of the figure, respectively, Z will be their point of equilibrium. Therefore the weight will be moved up the inclined plane KM by the force Γ, which moves it along a horizontal plane, *plus* the force Δ, which balances it when in position on the plane. Force Δ is thus proportional to the angle of inclination.

(marble rather than, say, water) and shape of the body to be moved (preferably spherical or cylindrical).[49] He also tackles the problem of how to move a body upwards along a plane perpendicular to the horizon: one needs first a force equivalent

[49] Hero, *Mech.* 1.20–21.

to the body, in order to balance it, and then a bit more force in order to move it. The procedure is analogous for a body moved up an inclined plane: Hero takes the body to be a cylinder, which he cuts across first with a perpendicular plane, in order to produce a circle, and then with another plane perpendicular both to the first plane and to the horizon. Although Hero does not construct this explicitly as a geometrical diagram, it is evident that the configuration of the *Mechanics* is the same as in Pappus' discussion (see fig. 3.2). In both cases a certain line (EΘ in the figure) is imagined as a balance, with the weight at one end and the counterbalancing force at the other, in the same ratio as the arms of the balance – in Hero's words, "thus we need a force equivalent to this difference [between the two parts] to keep [the body] where it is." A bit more force will be sufficient to move the weight along the inclined plane.[50]

The natural question at this point is: if Pappus knew of Hero's account (as he did), then he must have known that it takes only a minimal force to move a body along a horizontal plane, so why did he make the mistake he is accused of? Or, if he disagreed with Hero on this point, why did he not say so? The point is, I think, that Pappus presents the result in a different way from Hero. The problem of the inclined plane in the *Collection* is completely inscribed within a mathematical framework, not just for the demonstration itself, but for the propositions that precede and follow it. Physical considerations such as the preoccupation with smoothness that we find in Hero have been eliminated, and the importance of the amount of inclination of the plane is brought out: the angle of inclination is (I think) one of the initial data of the problem, and also one of the data on which the solution to the problem depends. It thus falls in line with Pappus' interests as we have described them. This is not to say that Pappus abstracts completely from the material circumstances: the connection with the real world of specific bodies moved up specific inclined planes is made by the numerical example appended to the

---

[50] *op. cit.* 1.23.

general demonstration. In any case, if we bring to the fore the fact that Pappus' proposition is a mathematical treatment of the problem, it becomes clear that it does not matter, from a *geometrical* point of view, how big or how infinitely small the force Γ is: it must still be there as a recognizable entity, as one term in the ratio; it still enters the consideration of forces that counterbalance the weight and then tip the scales, managing to move it.

Pappus' next task is again that of moving a weight by means of a force, but this time by actively intervening on the situation so as to employ a smaller force to move a greater weight. The problem is the perfect example of a case where the weaker vanquishes the stronger, that is, a quintessential mechanical problem, solved by means of a mechanical device. The issue and the device had again already been discussed at length in Hero's *Mechanics*, as Pappus explicitly recognizes. The order of discussion is different: in Hero (at least in the text as we have it) the machine comes first, then he considers how to move a weight along a horizontal plane, how to move a weight up a perpendicular plane, how to move a weight up an inclined plane. In the *Collection*, as we have seen, consideration of the natural situation (how to move a weight up an inclined plane in the absence of devices) comes before consideration of the mechanical situation. Without speculating too much, we could say that, while mechanics as it emerges from Hero's text is seen against a physical background – we are continuously reminded of the material nature of the objects, of their concrete characteristics and of the way they affect the reliability of the propositions Hero establishes – Pappus' mechanics (notwithstanding what is stated in the introduction) is made to operate within a mathematical universe of reference. Thus, one first examines the mathematical structure of the relation between weights and forces, and then transports it to a more real situation, adding flesh to the basic skeleton, as it were. That said, neither Hero nor Pappus polarizes the subjects of their inquiries – as we have already pointed out, Pappus is keen to emphasize the complementarity of mathematics and mechanics throughout book 8.

Pappus' device is a πολυσπάστον, i.e. a sysːem of inter-locking cog-wheels operated by means of a handle which is fitted to an endless screw.[51] Many details of the account, from the device itself to the theorems underlying its ɔonstruction, were contained in Hero's text, and Pappus does not fail to recognize it, declaring that the earlier work was "thoroughly clear."[52] He chooses not to reproduce some ɪesults which have already been proved by Archimedes (from a book on balances), Philo and Hero. Pappus also mentions that Archi-medes was allegedly the inventor of the πολυσπάστον, with reference to which he pronounced the famous words "Give me a place to stand on and I'll move the Earth." It could be that Pappus got the story at third or fourth haɪd, since the Doric dialect inflections found in another source have here disappeared.[53] The results that Pappus chose not to reproduce include the theorem about larger circles prevailing over smaller ones, a version of which is first found in the pseudo-Aristotle. There is no trace of the theorem in Archimedes' extant works, and I am inclined to think that Pappus had it from Philo or Hero.[54] When Pappus takes up Carpus' criti-cism, it would have been obvious for him to refer to an Ar-chimedean text of mechanics, had he been in possession of one. The fact that he finds less direct ways to deflect the

---

[51] The term πολυσπάστον is found in e.g. Plutarch, *Vita Marcelli* 14.7–8 (306); Vi-truvius, *De Arch.* 10.2.10 (*polyspaston*).

[52] 1060.4: πάνυ σαφῶς.

[53] 1060.3–4: δός μοί (φησι) ποῦ στῶ καὶ κινῶ τὴν γῆν. Simpliciːus, *In Aristotelis Physicam* VII 1110.4–6 has: πᾷ βῶ καὶ κινῶ τὰν γᾶν. It could bɛ that Simplicius' Doric inflexions are the result of a forgery. In Plutarch, *Vita Marcelli* 14.7–8 (306), Archimedes says that "if there were another world, and he could go to it, he could move this" (εἰ γῆν εἶχεν ἑτέραν, ἐκίνησεν ἂν ταύτην μεταβὰς εἰς ἐκείνην).

[54] 1068.19–23: ἀπεδείχθη γὰρ ἐν τῷ περὶ ζυγῶν Ἀρχιμήδους καὶ τɔῖς Φίλωνος καὶ ˚Ηρωνος μηχανικοῖς, ὅτι οἱ μείζονες κύκλοι κατακρατοῦσιν τῶν ἐλασσόνων κύκλων, ὅταν περὶ τὸ αὐτὸ κέντρον ἡ κύλισις αὐτῶν γίνηται ("it is proved in fact in the 'On balances' of Archimedes and the 'Mechanics' of Philo and oƒ Hero, that the greater circles prevail over the smaller ones when the motion cf both is around the same centre"). Our extant evidence only has the propositiɔn in the pseudo-Aristotle, *Mech.* 848b–849b and in Hero, *Mech.* 1.4–7. On the question whether Hero knew the pseudo-Aristotelian treatise, cf. De Gandt (1ɕ82), who thinks maybe; Micheli (1995), who thinks not. A large chunk of the second book of Hero's *Mechanics*, 2.34, is in the form of question-and-answer, and deals in many cases with exactly the same problems as the pseudo-Aristotle, buː could simply be an interpolation.

objection suggests that no direct counter-proof was available, and that Pappus did not *have* Archimedes' book on balances.

To get back to the *polyspaston*, the only significant difference between book 8 and Hero's treatment, and one which is explicitly underlined by Pappus, is that Hero chose 5:1 for the ratio between the diameter of a wheel and the diameter of the axis of the wheel, so that a weight of one thousand talents could be moved by a force of five talents.[55] As a matter of fact, Hero found that with his mechanism the weight could in principle be moved by a force of *four* talents, but, he added, the excess of one talent would have been lost anyway because of the wheels' resistance to turning. Pappus, on the other hand, postulates a ratio of 2:1, a weight of 160 talents and a force of 4 talents, which will be sufficient to balance the weight when suspended, so that adding a minimum extra force will move the weight. Why Pappus should choose to modify the existing account in a small numerical detail could be either a matter of difference at all costs, or consideration of the unfeasibility of a very large machine, or his decision to emphasize the numerical relations which underlie the actual functioning of the mechanism. In fact, after discussing other issues, he gets back to the geometrical foundations of the engine with the third programmatic problem, which is accompanied by a theorem on the ratio of circumferences to each other (already found in book 5) and the description, declaredly based on Apollonius, of a monostrophic spiral (a spiral that winds up on itself).[56] The spiral (again following Hero's lead) is to be imagined in terms of a screw, and the circles can be viewed as the mathematical armature of cogwheels: they constitute not so much a theoretical counterpart as a justification of the possibility itself of the mechanism, which, analogously to a geometrical problem, can only be constructed if all its constituent elements can be determined.

While a correct description of the machine depends on ge-

---

[55] Hero, *Mech.* 1.1 and again at 2.21.
[56] 1104.27–1106.25; the spiral at 1108.30 ff. Book 5 at 335.26 ff. Cf. the same curve in relation to a screw in Hero, *Mech.* 2.5 and the description of a monostrophic spiral in Proclus, *In Eucl.* 105.5 ff.

ometry, its actual functioning is linked to physical factors. In this sense, Pappus' solution can be seen as a mathematical skeleton which one has to flesh out with "concrete" circumstances. The actual functioning itself does not feed back into the mathematical description – rather, it counts as a sort of qualification of the geometrical solution that need not affect its correctness. The mathematical description expresses how the thing is from the point of view of its being an object with such-and-such a shape, i.e. a geometrical object: if the functioning does not occur in accordance with what is written down, it is because of a number of other factors, not because the description is in itself wrong.[57]

The second programmatic problem, that of the two mean proportionals, is tackled by Pappus in book 3 of the *Collection* as well, and we will discuss it in the next chapter. Here, it gives him an opportunity to make some points about the complementarity of geometry and mechanics.[58] The problem of the two mean proportionals, which is equivalent to that of the duplication of the cube, ought to be solved by means of conic sections, yet, because conic sections are difficult to draw in a plane, a different procedure is preferred which employs a moving ruler and is thus characterized as "mechanical."

Mechanics in a sense is above geometrical categories, because it includes geometry among the sciences belonging to its discursive part. Yet at the same time there are geometrical problems which are situated "outside mechanics."[59] In order to solve some of them – the duplication of the cube is a case at hand – geometrical procedures are at times not viable, and solutions must be found with the help of instruments (moving rulers or more complicated devices). On the other hand, geo-

---

[57] Similar sentiments in Hero, *Mech.* 1.13: "It does not cause any problem if one applies this statement to things which are perceptible by the senses; with things which are only thought of, however, it is even truer and righter" and 2.32: "Since it is not possible to human beings to produce [machines] with perfect smoothness and homogeneity, one must strengthen the forces against the friction of the machines ... so that one does not have a problem [because] our observation of the use of the instrument proves false something whose proof had been found to be right."

[58] 1070.1 ff.

[59] 1070.4: χωρὶς τῶν μηχανικῶν.

metrical reasoning can be deployed to determine the conditions of possibility and functioning of a mechanism, as in the case of the cog-wheel device.

The complex relationship between mechanics and geometry has a further subtlety in that there are problems to do with instruments which lie this time outside the domain of geometry and are thus, properly speaking, instrumental problems. An example in this category is the problem of determining the thickness of a cylinder whose bases have both been chopped off along irregular lines. Pappus says that this problem is put forward by the architects.[60] The solution is carried out by means of compasses with which one fixes several points on the surface of the cylinder (which we can visualize as a broken column). Outside the domain of geometry as it may be, the problem allows Pappus to launch into a number of related discussions involving the determination of several elements of an ellipse, given the position of some points on its circumference, or the distance and inclination of a sphere suspended above the horizon.

Further on, the relation between mechanics and geometry is once more reversed: Pappus constructs seven regular hexagons within a given circle. The procedure he uses is made so easy by (geometrical) analysis, he says, that it can dispense with the corresponding πεῖρα (experience, test).[61] The problem once again has an architectural background (Pappus includes it among "useful" instrumental problems): we have several examples of this type of geometrical pattern from mosaic floors.[62] We could take it that in this context πεῖρα would have indicated a check on the accuracy of the preliminary drawing, before starting the mosaic work itself, i.e. before starting to lay the *tesserae* into place.

---

[60] 1072.31–1074.2: Τὰ δ' ὀργανικὰ ἐν τοῖς μηχανικοῖς λεγόμενα προβλήματά [ἐστιν ὅτι] γίνεται τῆς γεωμετρικῆς ἐξουσίας ἀφαιρούμενα ("The so-called instrumental problems in mechanics are those generated at a remove from the domain of geometry"). The problem of the cylinder at 1074.3–1076.11.

[61] 1096.17 ff.

[62] Bertelli (1988), 39, has a second-century AD example from Tunisia. I thank A. Presas i Puig for drawing my attention to this illustration.

The interplay between geometrical analysis and mechanical procedures is therefore not univocally determined. In the *Collection* we have legitimate examples of all the possible combinations: purely geometrical, purely instrumental, mechanical constructions supported by deduction, mathematical solutions complemented by πεῖρα or by examples where the general proposition is applied to a set of specific numbers – a sort of mathematical equivalent of πεῖρα.[63]

The third programmatic problem is finally picked up for discussion, and explicitly linked to the πολυσπάστον – namely, the wheels whose teeth are to be determined are the same as the cog-wheels of the device. As we have said, Pappus complements the proposition with a theorem on the ratio between two circumferences expressed in function of their diameters. He also approaches the problem with the help of numbers, as we will see in more detail in chapter five, and includes an "instrumental" demonstration.[64] The debt to Hero is again acknowledged: "this indeed Hero proved in the Mechanics, and it is written down by us as well, so that nothing which is not included has to be looked for."[65] Nonetheless, neither the theorem on circles nor the twofold geometrical/numerical approach are in Pappus' source. We can also note that one of the terms used here for "tooth" in a cog-wheel is σκυτάλη, analogous to the σκυτάλιον of a document (AD 441) from Oxyrhynchus, which mentions repairs to the cog-wheel of a water-lifting device.[66]

Conclusion

As in the rest of the *Collection*, Pappus' attitude reflects a complex balance between a concern with his own agenda and

---

[63] The proposition on the centre of gravity of a trapezium at 1040.26–1044.15 contains a passage introduced thus: "if we imagine from the point of view of experience" (ἐὰν γὰρ ἀνὰ πεῖραν ἐπινοήσωμεν, 1042.11–12) where the reader has to imagine the weight of the trapezium all concentrated in one point.

[64] 1108.22–29.

[65] 1114.5–7: τοῦτο γὰρ Ἥρων ἀπέδειξεν ἐν τοῖς μηχανικοῖς, γραφήσεται δὲ καὶ ὑφ' ἡμῶν, ἵνα μηδὲν ἔξωθεν ἐπιζητῶμεν.

[66] Oleson (1984), 157, quoting *P. Mil.* 64 (*SB* 9503); see also Calderini (1920).

a respect for the weight of previous tradition. Having started the account with his own definition of mechanics, he uses it as a means of incorporating previous definitions, whose adequacy is, however, never put in question. He respects the classical themes of mechanics and reports canonical problems and myths, like the claim of Archimedes to be able to move the earth. On the other hand, he modifies transmitted results in accordance with traits that can be considered typical of the *Collection*: an emphasis on determining of the conditions under which a problem can be solved, the examination of subcases and (in two cases in book 8) the rehearsal of general theorems or problems by application to sets of specific numbers, usually chosen with an eye to facility of calculation.

As in traditional accounts of mechanics, utility has a prominent role. For instance, the duplication of the cube can be used to modify any solid according to any ratio, and therefore it is useful (χρήσιμος); the cog-wheel mechanism is also certainly useful, and several problems are explicitly taken from architectural practice. The theorems of mechanics are useful in another sense too: they are versatile, and can be applied in several situations. In this sense, a theorem can have a claim to usefulness without necessarily having an impact on actual reality. From this perspective, "useful" is also what helps towards the causal enquiry that Pappus identifies with mechanics: the cog-wheel has not only the use of moving weights, but also an epistemic value, in that it enlarges our knowledge. In my opinion, this is the light in which we should read Eutocius' much later reference to Archimedes' *Measurement of the Circle* as "necessary for the uses of life."[67] The practical implications of Archimedes' discovery were certainly not missed by the ancients, and they do colour the meaning of the term "useful" here, but then again so do the potentialities of the work for a wide range of mathematical problems.

---

[67] *In Dim. Circ.* 228.21: πρὸς τὰς τοῦ βίου χρείας ἀναγκαῖον. Eutocius says that he is quoting from Heraclides' *Life of Archimedes*.

Indeed, the semantic range of terms such as χρεία or *utilitas* in the Greco-Roman world overlaps, but is not completely mapped by, our own "utility" or "usefulness." As we have had occasion to remark, *utilitas* was frequently invoked in Roman laws as a justification of what was being done in the name of the common good. Many writers of technical treatises also asserted that their discoveries were useful: for instance, the anonymous author of the fourth-century *De rebus bellicis* made that claim for many of his war machines, some of which, it has been observed, far from being useful, are simply not feasible. The problem is, did the author know that some of his machines could have never been built? And if so, why did he describe them at all, let alone claim that they were useful? I do not have an answer to these questions, which indeed could be asked for a number of other machine inventors, up until at least Leonardo da Vinci. I suspect that at the time some of these texts were written, "feasible" was not as indispensable a constituent of "useful" as it is now. It was a good thing for an author to show that he could think up something of practical use, and it was a good point for him to claim that he had direct experience of things; at the same time, producing something extraordinary was appreciated even if it remained confined to the written page or to a colourful illustration, perhaps because it represented, in a sense and in any case, a material display of power.

The point is thus less about real applicability than about the *rhetoric* of utility – the point is that there was one, and it was alive and kicking in late antiquity. Proclus reports the opinion of some detractors of mathematics (with whom he disagrees, which makes the testimony all the more valuable) thus:

Some people declare that the empirical sciences concerned with sense objects are more useful than the general theorems of mathematics. Land-surveying, they say, is more useful than geometry, the arithmetic of the many more than the foundation of theorems, and navigational astronomy more than general astronomy ... so we shall agree that the empirical sciences, not the theories of the mathematicians, contribute most to human life and actions. Those who are ignorant of reasonings but practised in dealing with particular problems are far and away superior in meeting human needs to those who have spent their time in the schools pursuing theory alone.[68]

Later, Cassiodorus compared the useful art and positive civic qualities of the land-surveyor with the empty schoolrooms of arithmetic teachers and the limited appeal of astronomy. Again, the anonymous author of the *De rebus bellicis* commented: "In this connection one has always to examine what a person means rather than what he says; for it is patent to everybody that not the greatest nobility, nor abundance of resources nor the powers rooted in the legal courts or the eloquence acquired with letters has come up to the advantages of the arts."[69] The utility of the arts may not have been as patent to everybody as the author makes it out to be, but it was indeed the case, as we have seen in chapter one, that the government actively promoted a continuity of practice and teaching for mechanical disciplines such as architecture.

The public "clout" of mechanics came from its official recognition as beneficial to the community. Wonder-working, on the other hand, was out of favour, given the bad press magicians, astrologers and other *mathematici* were receiving, and given that in some cases the laws we have mentioned in the first chapter were actually enforced for political reasons.[70] In the event, the wondrous aspects of mechanics are understated in the *Collection*, while the officially favoured value, utility, takes central place.

---

[68] Proclus, *In Eucl.* 25.12–26.9: οἱ δὲ χρησιμωτέρας τὰς τῶν αἰσθητῶν ἐμπειρίας ἀποφαίνονται τῶν ἐν αὐτῇ καθόλου θεωρουμένων, οἷον γεωδεσίαν γεωμετρίας, καὶ τὴν τῶν πολλῶν ἀριθμητικὴν τῆς ἐν θεωρήμασιν ὑφεστώσης, καὶ τὴν ναυτικὴν ἀστρολογίαν τῆς καθόλου δεικνυούσης ... ὥστε καὶ πρὸς τὸν βίον τὸν ἀνθρώπινον καὶ τὰς πράξεις οὐ τὰς γνωστικὰς τῶν μαθηματικῶν, ἀλλὰ τὰς ἐμπειρικὰς συντελεῖν ὁμολογήσομεν. οἱ γὰρ ἀγνοοῦντες μὲν τοὺς λόγους, γεγυμνασμένοι δὲ περὶ τὴν ἐν τοῖς καθ᾽ ἕκαστα πεῖραν ὅλῳ καὶ παντὶ διαφέρουσι πρὸς τὰς ἀνθρωπικὰς χρείας τῶν περὶ τὴν θεωρίαν μόνην ἐσχολακότων.

[69] *De rebus bellicis* Pref. 6: "In qua re est considerare semper quid unusquisque magis sentiat quam loquatur; constat enim apud omnes quod nec summa nobilitas nec opum affluentia aut subnixae tribunalibus potestates aut eloquentia litteris acquisita consecuta est utilitates artium." (Trans. here and henceforth Thompson, with modifications.)

[70] Averil Cameron (1993), 164, has examples from Ammianus Marcellinus: see especially *Res Gestae* 29.2 (set around AD 371).

Again, Hero's treatise on war engines, as I have argued elsewhere, addressed a concern for security which could be situated historically in the eventful circumstances of Egypt and Palestine in the second half of the first century AD. A renewed concern for security is undoubtedly expressed in the *De rebus bellicis*, which identifies a menacing "outside" where an alien and hostile environment and alien and hostile populations work together to threaten the Empire.

> Above all it must be recognized that the Roman Empire is pressed upon by nations howling around it everywhere, and treacherous barbarians, protected by their natural environment, are assailing every frontier. For usually these nations are either sheltered by forests or climb mountains or are defended by snow and ice, some are nomadic and are protected by deserts and the burning sun. There are some who are defended by marshes and rivers.[71]

War technology is not represented in the *Collection*, but just as belopoietics and wonder-working can be seen as expressions of a strong link between mechanics and political power, then it makes sense to say that book 8 hints at a different relation between mechanics and power, or, in broader terms, science and power. Pappus' interlocutors never seem to be, not even implicitly, direct suppliers of patronage or individual repositories of political power, as was the case, for instance, with Biton, Vitruvius, Apollodorus and the author of *De rebus bellicis*, who all addressed their works, all dealing in one way or the other with war technology, to kings or emperors.[72] Rather, the emphasis Pappus gives to some aspects of mechanics is political to the extent that a concern for utility is a concern for public utility, and awareness of the past and respect for the tradition of a discipline were generally shared values. In other words, the shift in Pappus away from mechanics as the spectacle of individual-centred power, towards a more sober exercise in understanding reality also signifies a different politics of knowledge. In a parallel to the distinction made by the author of *De rebus bellicis* between himself as a

---

[71] *De rebus bellicis* 6.1–4. See also Tomei (1982); Cuomo (forthcoming).
[72] Respectively: King Attalus of Pergamon, Augustus, Hadrian and Valentinian I and Valens, or Constans II with one of his Caesars; see Schürmann (1991).

private citizen and the repositories of public power he is addressing, the persuasive strategies of book 8, the ways in which legitimation is sought by Pappus suggest private citizens rather than emperors. Although the values exemplified by the use of the past and by the importance of utility were indeed promoted by the government, here it is as values shared by local groups that Pappus appeals to them.

Let us re-read Pappus' programme and his many side-comments on the relative standing of geometrical and mechanical procedures in a "late ancient" light. In a field like mechanics, where in many cases one could actually do without geometry, Pappus is keen to emphasize that the two disciplines are beneficial to each other; they complement one another in a two-way, "democratic" manner. His defence of the indissolubility of mechanics and mathematics is not only a tribute to authoritative traditions, but also a claim to a wider source of legitimation for mathematicians. He stresses the role of geometry as *both* educationally or culturally useful (embodied by the scholar as mouthpiece of the tradition) *and* materially useful (personified by the able architect/engineer/ mechanician) – in fact, he stresses that the two aspects essentially cannot be separated. Hence, in my opinion, the pictures we find in the *Collection* of the globality of mechanical education, which at least on an ideal level must encompass arts and sciences, and of harmony between mathematical and mechanical demonstrations. Globality, harmony and utility, especially if supported by a long-standing tradition, were, as we have seen, key values at Pappus' time.

CHAPTER 4

ALTARS AND STRANGE CURVES

In the last two chapters we have seen how Pappus juxtaposes his own form of knowledge to other ways of knowing – on the one hand, he marks out the difference between mathematicians and philosophers; on the other hand, he states the complementarity of mechanics and mathematics. In both cases Pappus' attitude reflects issues which characterized perceptions of mathematics and, in general, of knowledge at the time. In both cases, he uses the mathematical traditions before him to empower himself and legitimate what he is doing. In what follows I will look at how Pappus deals with different ways of doing mathematics, i.e. how he marks out his territory within the mathematical field itself. Once again, his approach relates to the past on the one hand and to the present on the other.

4.1 Doing it right (or otherwise)

In book 3 Pappus tackles the problem of the two mean proportionals, or duplication of the cube, the so-called Delian problem, now also known as one of the three "classical" problems of Greek geometry.[1] The book is addressed to the woman Pandrosion, who from the context emerges as a teacher of mathematics – Pappus refers to some people who make out that they are her pupils.[2] One of these people, who

---

[1] 30.3–68.16. The other two classical problems are the quadrature of the circle and the trisection of the angle; cf. Knorr (1986a). The reduction of cube-duplication to the problem of finding two mean proportionals is attributed to Hippocrates of Chios, cf. DK 42A4 (Eratosthenes *apud* Eutocius, *In Archimedis sphaeram et cylindrum* 88.17–23); cf. also Saito (1995).

[2] Jones (1986a), 4n8: "Pappus's Pandrosion has suffered strange indignities from Pappus's editors: in Commandino's Latin translation her name vanishes, leaving

remains unnamed, had sent Pappus the sketch of a solution to the problem of the two mean proportionals, i.e. a construction probably with a setting-out (description of the construction) but without a full proof, plus two other problems.[3] We only have Pappus' version and are therefore unable to give any independent characterization of the anonymous geometer's text or original intentions. Pappus of course is hardly interested in painting a faithful picture of the other's contribution; rather, it is a good opportunity for him to start a statement of the correct criteria for a valid solution to the problem with an exposition of other people's mistakes. In other words, the construction is considered by Pappus a badly flawed attempt.[4]

Particular attention ought to be paid to Pappus' introductory remarks. The unnamed geometer, he says, has claimed to be able to determine two mean proportionals to two given lines by means of "planar investigation."[5] No proof (ἀπόδειξις) is supplied, although we are told that the geometer has in fact promised one.[6] What Pappus does have is a construction (κατασκευή), which he has been asked to examine and respond to.[7] As a result, he raises three objections: first of all, that the position of the point Φ is insufficiently determined; secondly, that the demonstration is circular and, finally, that

---

the absurdity of the polite epithet κρατίστη being treated as a name, "Cratiste"; while for no good reason Hultsch alters the text to make the name masculine." The alleged pupils at 30.17–18: [τίνες] τῶν τὰ μαθήματα προσποιουμένων εἰδέναι διὰ σοῦ.

[3] Omitting to send a proof was quite common practice in antiquity: cf. Archimedes, *Sph. Cyl.* I & II, Introd.; *Conoides et Sphaeroides* Introd.; *Spir.* Introd.; *Meth.* Introd.; Apollonius, *Conicae* IV Pref. I thank R. Netz for his comments on this point. The other two problems will not be discussed in detail: 68.17 ff. is about finding an arithmetic, a geometric and a harmonic mean within the same semicircle (on this see M. Brown (1975)); 104.14 ff. states that, given a right-angled triangle, it is possible to construct a certain triangle internal to the first one and such that the sum of two of its sides will be greater than the sum of the two sides of the triangle containing them.

[4] Pappus' attitude to the other two problems is more lenient: in the proposition about means, he says that the author has not specified several things, and does not really know how to recognize a harmonic mean, 68.31–70.4; in the proposition about the triangles, Pappus puts forth a more correct enunciation, 104.24.

[5] 30.25 f.: λαβεῖν ἔφασκεν εἰδέναι δι' ἐπιπέδου θεωρίας.

[6] 34.1–8.

[7] 30.25–32.2: ἐπισκεψαμένους ἀποκρίνασθαι.

"planar" procedures (i.e. straight-line-and-circle constructions; we shall look at Pappus' definitions later) are improperly applied to a "solid" problem.

The position of Φ varies according to the ratio $K\Theta{:}\Theta P$, which is equal to the ratio $BE{:}EA$ – but the author, Pappus says, has failed to specify that, and has not indicated that the ratio $BE{:}EA$ is given. On today's criteria, this first objection may seem superfluous,[8] yet Pappus uses quite harsh terms: the difficulty (ἀπορία) stemming from the construction reveals its author's "responsibility", the fact that he is the "reason" for the ἀπορία itself. I think that the term αἰτία here retains its double connotation of "cause" – the fallacies of the construction are due to incorrect reasoning – and, in a more personal tone, "responsibility", "blame" – the incorrect reasoning is somebody's.[9] Pappus also accuses the other geometer of having overlooked the consequences of not giving the ratio $BE{:}EA$ at the outset – this is characterized as an attempt to investigate the impossible[10] and again alluded to as the author's responsibility.[11]

The passage strongly implies that, from Pappus' point of view, the non-determination (the verb is ὁρισθῆναι) of the position of Φ, a problem of ambiguity in the figure, amounts to impossibility of the construction, and the impossibility of the construction has the natural consequence (given also the lack of a proof) that the entire solution to the problem is false.[12]

But, somewhat surprisingly, Pappus seems prepared to make some concessions. Once the initial ratio $BE{:}EA$ *is* given, one should be able to rectify the first mistake by simply calculating how the position of Φ varies with the variation of the initial ratio. This is exactly what Pappus proceeds to do: he substitutes increasing numerical values for the ratio $K\Theta{:}\Theta P$

---

[8] This is the opinion of Knorr (1989), 69.

[9] 34.13–15: τῆς δὲ τοιαύτης ἀπορίας παρὰ τὴν αὐτοῦ αἰτίαν ἐπακολουθούσης ἐνεφάνισεν ἑαυτὸν μηδὲ τοῦτο συνιδόντα τὸ ἀκόλουθον.

[10] 34.18 f.: πειρᾶται ζητεῖν τὸ ἀδύνατον.

[11] 38.11: δι' ἣν εἴπομεν αἰτίαν.

[12] 34.15: ἀδυνάτου γὰρ ὄντος ὁρισθῆναι τὸ τῆς τομῆς σημεῖον. Cf. Proclus, *In Eucl.* 204.5–7: "When, then, do we say the exposition is lacking? When the enunciation contains no statement of what is given."

and determines the position of Φ in each of those cases. He then states the general conditions for the position of the point on the basis of the initial ratio and arranges the results into four lemmas, which are incorporated into the text.[13] All the elements of the construction should now be determined. Yet Pappus insists that, no matter where Φ is situated, the construction remains fallacious.[14]

Pappus' second objection is that the statement KΘ:ΘΣ = ΘΣ:ΘT = ΘT:ΘP is circular: ΘΣ and ΘT are mean proportionals between KΘ and ΘP, therefore one presupposes what one is trying to find. Now, if we take a step back, i.e. to the first report of the construction, we see that nowhere is it said that KΘ:ΘΣ = ΣΘ:ΘT = ΘT:ΘP. The last term of the proportion here is ΘΦ – the charge of circular reasoning still holds, but why should Pappus have substituted the term ΘP for ΘΦ?[15] Let us leave this point aside for the moment. Again, Pappus makes some concessions – in this case, he takes it that the positions of Φ, B′ and Θ′ *are* given. He then substitutes numerical values for the ratio KΘ:ΘP and for ΘK and works out whatever data of the construction can be calculated on the basis of them.

The anonymous geometer's construction uses a sort of approximation procedure by means of iteration, so the construction repeats itself, as it were, in three cases (going from the right to the left in fig. 4.1).[16] Pappus recognizes that the geometer has tried to postulate Φ as the starting point of the iteration process, whereby point B′ in the second case would tend to coincide with point ϛ and point Θ′ would tend even more to coincide with point ϡ. Yet, he is both misguided and misleading his readers, and gets further blame for not daring to state his alleged intention explicitly.[17] In fact, no matter how many iterations the process undergoes, no matter how close any pair of points, be they Φ and P, B′ and ϛ, Θ′ and ϡ,

[13] 48.19–52.30.
[14] 38.14–15.
[15] 32.12–14. Hultsch blames this point on the "obscuritas" of the text, 39, note **.
[16] See Pendlebury (1873).
[17] 40.16–21 and 40.12, respectively.

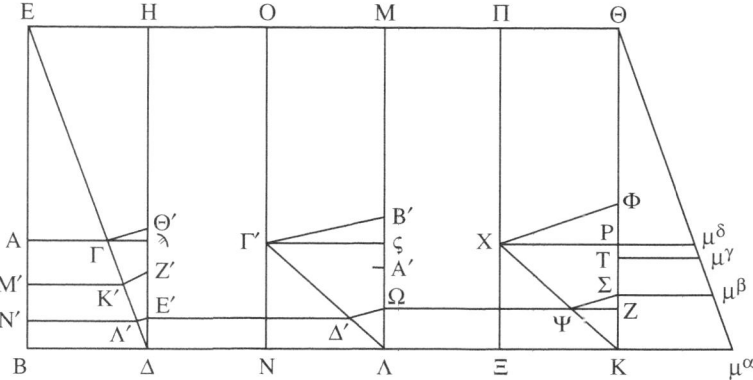

Fig. 4.1. We want to find two mean proportionals between ΑΓ and ΒΔ. The diagram is constructed so that AB = BΔ = ΔN = NΛ = ΛΞ = ΞK = Kμ$^\alpha$ = EH = HO = OM = MΠ = ΠΘ. Also, we take KP = AB and PΣ = ΣK and, again in the right-end corner of the diagram, we take T and Φ such that KΘ:ΘΣ = ΣΘ:ΘT = ΘT:ΘΦ. Next, we take ΧΞ = AB, join X and Φ and draw ΨΣ parallel to it, and draw the line ΩΨ parallel to BΛK. Again, we take Γ'N = AB, then A' and B' such that ΛM:MΩ = ΩM:MA' = MA':MB', join Γ' and B', draw the ΩΔ' parallel to it and the Δ'E' parallel to the BΛK. Finally, we take ϡΔ = AB, then Z' and Θ' such that ΔH:HE' = HE':HZ' = Z'H:HΘ', join ΓΘ' and draw K'Z' and Λ'E', both parallel tc it. The two lines K'M' and Λ'N', parallel to BΛK, will be the two mean proportionals between ΑΓ and ΒΔ.

get to each other, they still will not coincide – the approximation is bound never to come to an end. Pappus reconstructs a fourth instance $(\mu^\alpha, \mu^\delta, \mu^\gamma, \mu^\beta)$ of the iteration procedure so that, in the diagram, its position is right next to the first instance (Φ, P, T, Σ, Z). He then proceeds to show that the same fallacious consequences inevitably arise in that extra case as in the other ones. The points Φ, B', Θ' or indeed μ^β or μ^γ are given only on the circular assumption that we have already determined two mean proportionals between two lines – the point is not one of more or less close approximation, but of unwarranted starting-points. With the fourth instance of the iteration procedure, Pappus wants to lead the reader to conclude, by simple inductive reasoning from one part of the diagram to the next, that the entire iteration process does not work: what is false in one instance is false in the others as well. In my opinion, Pappus' remark that the construction deceives readers "by means of more and more"[18] is also to be read in this light: that is, as pointing out that the sense of persuasion generated by the iteration process is due to sheer repetition of instances. Pappus' objection counteracts it in that it induces persuasion of the contrary on similar lines, i.e. going through a number of instances and showing that the construction does *not* work.

I think (and we are back to the question left aside) that it is on his own reconstruction of the other author's intentions that Pappus feels free to criticize him for having assumed that ΚΘ:ΘΣ = ΘΣ:ΘΤ = ΘΤ:ΘΡ, and not ΘΦ. The pair of points Φ and Ρ are the first which are meant progressively to coincide. Taking their congruence as granted only to deny it later, the way Pappus does, can be seen as a rhetorical move somehow similar to a *reductio ad absurdum*: even when treated as if they were valid, the adversary's assertions lead to inconsistency. The author's planar procedures, in fact, were doomed from the start, because the problem is solid.[19] We are thus at the third of Pappus' objections.

The correct geometrical determination of the construction, as distinguished from just greater and greater approximation, can only be achieved by methods adequate to the nature of the problem. The anonymous geometer not only fails to produce a valid solution, he also exposes his ignorance or misunderstanding of the essential distinction between planar, solid and linear categories and of the methodological prescriptions related to it. At the conclusion of the passage Pappus strategically invites his readers to test and examine for themselves what he says with regard to the fallacious solution, because they are supplied with all the evidence they need.[20]

[18] 40.17: διὰ πλειόνων.
[19] 40.10: στερεὸν γάρ ἐστιν τῇ φύσει. On this see Étienne & Roels (1986).
[20] 52.31–54.3: Ἃ μὲν οὖν ἔδει με προειπεῖν ἐστιν ταῦτα, παρεὶς δὲ κρίνειν σοί τε καὶ τοῖς ἐν γεωμετρίᾳ γεγυμνασμένοις τὰ ὑπ᾽ ἐκείνου προγραφέντα περὶ τῆς κατασκευῆς καὶ τὰ ὑφ᾽ ἡμῶν ἐπενεχθέντα, καλῶς ἔχειν ἡγοῦμαι καὶ τὰ δόξαντα τοῖς ἀρχαίοις περὶ τοῦ προειρημένου προβλήματος ἐκθέσθαι.

On a less technical level, this long passage gives us an insight into the mathematical milieu of Pappus' time. The author had sent the construction to Pappus,[21] but probably not only to him, because others were acquainted with it as well. We are told that it had caused quite a stir and that many people, not necessarily experts in the field – they are identified as friends of the philosopher Hierius – had set out to examine it. It is not known how successful their efforts were: at some point Pappus decided to wait no longer for the author to publish the announced proof, and, taking advantage of the fact that his opinion had been asked in the first place, pronounced the account "not as it ought to be, but inexpert."[22]

References to what we may term unprofessionalism occur throughout the text: the unnamed geometer fails to realize that his construction leads to an ἀπορία; he tries to investigate the impossible, and expects his reader to do the same; he notices the circular reasoning, and not only does he not have the nerve to carry his assumptions to their furthermost consequences – the way Pappus does – but he effectively deceives the reader, while at the same time deceiving himself. Last but not least, he ignores the previous tradition of research.[23] Pappus, on the other hand, presents himself as a paradigm of fair dealing with an adversary: he devotes plenty of space to reporting the other's construction and, while detecting its fallacies, temporarily assumes the construction to be valid in order to develop it further. It thus appears as though he really did with it the best he could.

The criticism of the anonymous construction can be seen as a carefully engineered statement: of Pappus' mathematical ability, of the importance of certain topics which are at the heart of his practice (the requirements set on the solution to a

---

[21] 34.1–8. Cf. 30.24–32.3.

[22] 34.7–8: ὡς οὐ δεόντως, ἀλλ' ἀπείρως ἐχρήσατο τῇ κατασκευῇ. There may be a play on the word ἀπείρως, which means both "in an inexpert way" and "in an endless way" (with reference to the approximation procedure). I thank G.E.R. Lloyd for the suggestion.

[23] 34.11–15; 34.18–19; 40.16–19; 44.18–20 and 48.15–16, respectively.

problem; the issue of indeterminacy; the distinction between solid and planar problems), of the inseparability of expertise and more general virtues. In the next part of book 3, Pappus selects some "good" solutions to the problem of the two mean proportionals from among the wealth of material handed down to him, and caps everything with his own contribution.

4.2 Fishing in the stream of tradition

According to some of our earliest reports, whose common source seems to be Eratosthenes (end of third/beginning of second century BC), the problem of the two mean proportionals arose in the context of architectural practice. The god at Delos had requested a new altar twice as big as the extant one and, like the first one, in the shape of a cube, but the architects employed for the task simply doubled the side of the cubic altar, thus producing a second cube *eight* times as large as the first one.

At a loss for an answer to the puzzle, a delegation of Delians went and asked Plato for a solution, and the philosopher, after redirecting them for a more concrete answer to mathematical experts such as Eudoxus of Cnidos or Helicon of Cyzicus, revealed that the god's real purpose had been to inspire greater interest in the Muses, in mathematics and in other peaceful pursuits, which had long been neglected. Plato is also depicted by Plutarch as reproaching "those around Eudoxus, Archytas and Menaechmus" for trying to transform the Delian problem into a mechanical construction. An already mentioned passage of the *Vitae Parallelae* has Plato put on Eudoxus and Archytas the even greater blame of having tackled not one, but many a problem with sensory and instrumental examples (παρα-δείγματα) – the two geometers are presented as more or less single-handedly responsible for the debasement of geometry.

Eudoxus and Archytas . . . embellished geometry with its subtleties, and gave to problems incapable of proof by word and diagram a support derived from mechanical illustrations that were patent to the senses. For instance, in solving the problem of finding two mean proportional lines, a necessary requisite for many geometrical figures, both mathematicians had recourse to mechanical arrangements ... But Plato was incensed at this ...[24]

Plato may have been incensed, but, if we look at the mathematical traditions on the problem, we find that the duplication of the cube, far from being the epitome of pure mathematics, was quite generally associated with applied mathematics. It appeared in mechanical treatises (Philo, Hero),[25] and its practical applications were emphasized (Eratosthenes, Hero);[26] authors criticized their predecessors because of the cumbersomeness of their procedures (Eratosthenes, Nicomedes).[27] Indeed, all but one solution handed down to us from antiquity imply either the use of instruments other than ruler and compasses or the application of curves such as could only be drawn by point-wise procedures or by means of, again, instruments other than ruler and compasses. Moreover, instead of representing a means to incite men to peaceful pursuits, the duplication of the cube was in fact closely linked with the construction of war-engines, and it is in a treatise on catapults (Philo's) that we find its earliest first-hand complete solution.[28]

Whatever its historical origin, then, the problem over the years acquired various and contrasting meanings which overrode its mathematical significance and transformed it into an object of interest for philosophers as well as for mathematicians. The Hierius mentioned by Pappus is just the most evident example in this sense.[29]

The three solutions to the problem of the two mean proportionals reported by Pappus are not the only ones he knew

[24] See the story in Eratosthenes *apud* Theon Smyrnaeus, *Ad Legendum Platonem* 2.3–12; Plutarch, *Moralia* 386e; 579a–d and 718e–f and *Vita Marcelli* 14.5 (305); Eratosthenes *apud* Eutocius, *In Sph. Cyl.* 88.3–96.27. For the hypothesis that Plutarch's source is Eratosthenes' *Platonicus*, cf. Knorr (1986a), 3–4.

[25] Philo, *Belop.* 52; Hero, *Mech.* 1.11.

[26] Eratosthenes *ap.* Eutocius, *In Sph. Cyl.* 90.14–27; Hero, *Belop.* 114–115.

[27] *Ap.* Eutocius *In Sph. Cyl.* 90.4–11 and 98.5–7, respectively.

[28] Philo, *Belop.* 52.

[29] Proclus, *In Timaeum* 3.30.8–36.19, discussing Plato's *Timaeus* 32a7–b3 (on the ratios governing the cosmos), brings up the duplication of the cube and reports the views of his teacher Syrianus. Of the other two problems proposed by the anonymous author in book 3, that on means receives a parallel treatment in Iamblichus, *In Nicom.* 100.15 ff., while the "paradoxical" proposition on triangles is included in Proclus, *In Eucl.* 327.8 ff.

of: he mentions Philo's solution, contained in this latter's *Mechanica*,[30] and hints rather vaguely at other constructions, or attempts at constructions, "by means of conic sections."[31] Apollonius is mentioned, but Hultsch suspects an interpolation and, for once, he seems to have good reasons: the passage is abrupt and apparently meant to correct a blunder. Pappus has just said that the solutions by Eratosthenes, Philo and Hero were arrived at only instrumentally (μόνον ὀργανικῶς), and the supposed interpolator hastens to add (broadly put): well, of course, there are other proofs, by Apollonius via conic sections and Aristaeus via solid *loci*, but none of them employs planar methods properly speaking.[32] Pappus may also have known of Sporus' proof – one can even take the view that it inspired his own, as we shall see later.[33]

That several more solutions were known in antiquity is clear from Eutocius, who reports a number of them prior in time to Pappus, but not contained in the *Collection*. On the face of it, it would seem as if Eutocius had many more texts

---

[30] 56.1. Philo's solution is in *Belop.* 52 and in Eutocius' anthology, *In Sph. Cyl.* 60.28–64.14. Philo is also mentioned in the *Collection* in book 8 at 1068.19–23.

[31] 54.25. Cf. also 954.10–958.15 and Jones (1986a), II 488. Eutocius reports two proofs employing conic sections: Menaechmus' (*In Sph. Cyl.* 78.13 ff.) and the one immediately after, labelled "In another way"("Άλλως), *ibid.* 82.1 ff. Toomer (1976), 169–70, has identified this latter as Diocles', but see *contra* Knorr (1982b), 5–6.

[32] 56.4–5 and 56.6–7, respectively (οὐδεὶς δὲ διὰ τῶν ἰδίως ἐπιπέδων καλουμένων). While the proof ascribed to Apollonius by Eutocius (*In Sph. Cyl.* 64.15 ff.) makes no use of conic sections, Apollonius and Aristaeus are associated elsewhere in the *Collection*, 634.9; 672.18–24; 676.25–678.15, thus providing the assumed interpolator with grounds for the link. Cf. also Thaer (1940).

[33] Sporus' proof in Eutocius, *In Sph. Cyl.* 76.1 ff. The question whether Pappus knew of Archytas' proof, which to date is only found in Eutocius, is more complex. Eutocius' source (*op. cit.* 98.18–102.19) is declaredly Eudemus, although it has been argued (convincingly, I think) that in the sixth century AD Eudemus' *History of Geometry* was no longer available, and was only known indirectly, maybe through Sporus' Κηρία; cf. Tannery (1882); Neuenschwander (1974). In either case, Pappus could have been aware of Archytas through Sporus or directly from Eudemus or from an Eudemian source, which he mentions in his *Commentary on Euclid* 63. The name of Archytas was associated with the problem of the two mean proportionals in quite a general way, even in cases when the features of his proof do not seem to have been known in detail – he is mentioned by Vitruvius (*De Arch.* IX Pref. 14) and Plutarch (*Vita Marcelli* 14.5 and *Moralia* 718e–f) and, most notably, by Eratosthenes (*ap.* Eutocius, *In Sph. Cyl.* 90.6). Pappus had access to this latter source, from which he may have learnt not only of Archytas', but also of Menaechmus' solution, which could be identified as the object of the reference in the *Collection* to solutions via conics.

available than Pappus, but somehow I think that unlikely. Eutocius is nearly two centuries later; although he must have had access to a good library in Constantinople, if indeed he was based there, the same must have been true of Pappus in Alexandria. The difference in my opinion, is rather, in the selection operated on the sources. On the whole, I think that the extant evidence enables us to say that Pappus did not just report everything that was available to him, but rather that he *chose* some solutions. His selection was informed by criteria which I will try to identify in the next sections.

Eratosthenes

Eutocius' report of Eratosthenes' solution is more complete than Pappus'. It contains a dedicatory letter to King Ptolemy (identified with Ptolemy III Euergetes, 284–221 BC); a brief history of the problem, which credits Archytas, Eudoxus and Menaechmus with contributions; a proof complete with instructions for the use of an instrument, the mesolabe (literally, "mean-taker"); a slightly different version of the proof, and finally a poem with which Eratosthenes offers the mesolabe to the gods:

> If you plan, of a small cube, its double to fashion,
> Or – good sir – any solid to change to another
> In nature: it's yours. You can measure, as well:
> Be it byre, or corn-pit, or the space of a deep,
> Hollow well. . . .
> By these tablets, indeed, you may easily fashion –
> With a small base to start with – even thousands of means.
> O Ptolemy, happy! Father, as youthful as son:
> You have bestowed all that is dear to the muses
> And to kings. In the future – O Zeus! – may *you* give him,
> From your hand, this, as well: a sceptre.
> May it all come to pass. And may he who looks say:
> "Eratosthenes, of Cyrene, set up this dedication."[34]

The functioning of the mesolabe is described in detail by Pappus; he specifies how to slide the plates inside a groove

---

[34] Eutocius, *In Sph. Cyl.* 96.10–27, trans. in Netz (forthcoming c).

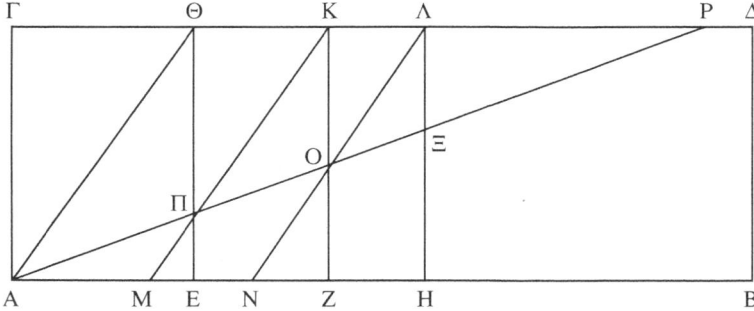

Fig. 4.2. ΑΒΓΔ is the frame of the mesolabe, the triangular plate ΑΕΘ is fixed, the plates ΜΖΚ and ΝΗΛ can slide along the frame. The plates are all equal triangles. Ξ is the mid-point of ΛΗ. The plates are moved until Α and Ξ are on the same line as Π and Ο. It will result that ΠΘ and ΟΚ are mean proportionals between ΑΓ and ΛΞ.

along the contouring frame so that they are arranged in a certain position.

At the end of the demonstration, Pappus points out that the aim of the construction is "to produce a cube double a cube," and that one can use the same method to find whatever multiple of the cube one wants, not just its double.[35] Among the solutions reported in the *Collection*, Pappus' and Eratosthenes' are the only ones to state the problem as one of cube-duplication.

Although the procedures in Pappus and Eutocius are similar (both employ sliding plates and appeal to the properties of proportions), they differ in their phrasing, especially the terms used to describe the mesolabe, and in the shape of the plates.[36] Moreover, in Eutocius' report, Eratosthenes makes some strong claims: earlier constructions, he says, however

---

[35] 58.5; 58.18–20, respectively.

[36] The device is called a πλινθίον (rectangular frame or box) in Eutocius and a πλινθίον πεπηγός (*rigid* rectangular frame or box) in Pappus; then again we have three πινακίσκοι or πίνακες (rectangular tablets) which move ἐν χολέδραις (in grooves) in Eutocius, whereas triangles move in a σωλήν (again, groove) or an ὀχετός (channel) in Pappus: *In Sph. Cyl.* 92.25 ff. and *Collection* 56.18, 23; 58.4 respectively. The specification "rigid rectangular frame" may be due to the fact that πλινθίον in book 3 is also used for "table" of means, 100.22, 26, 29; 104.3, 13.

rigorously demonstrated, were not apt for any practical purpose. His own method, on the contrary, can be used in a number of situations, from agriculture to military engineering, and with the same procedure one can find not only two, but three or more mean proportionals to two given lines, and, under certain conditions, the original cube can be converted into other solids.

At least some of the discrepancies could be put down to the use of different sources. In the past there have been doubts as to the authenticity of Eutocius' report, which is now thought to draw on Eratosthenes' *Platonicus*. Pappus' source is not known. The *Collection* says that mechanical solutions to the duplication of the cube are to be found "in the mesolabe of Eratosthenes and in the mechanics of Philo and Hero."[37] Since the two latter texts are actually entitled *Mechanics*, the obvious inference would be that Eratosthenes' proof was contained in a book called *On the mesolabe*. Other possible candidates are, again, the *Platonicus* or Eratosthenes' book *On Means*, which, along with Euclid's *Data*, Apollonius' *Conics* and others, was included in the so-called *Treasure of analysis*, a collective source employed for book 7 of the *Collection*.

In conclusion, then, we do not know anything with certainty about Pappus' and Eutocius' sources. It is at least likely, I think, that Pappus had access to the same sources for Eratosthenes as Eutocius, but decided to include only the proof and leave out the rest, in particular to leave out the history of the problem. What for Eutocius may have been well worth reporting would maybe have been redundant in the economy of Pappus' treatise, which is focused on showing how the problem is solved correctly (rather than incorrectly, as exemplified by the anonymous geometer). Eratosthenes' claim that his own solution was better than the previous ones would hardly have fitted in an account, like Pappus', where the author is basically making that very claim for himself.

[37] 54.31–56.1.

If we take Eutocius' report as an example of the type of source that would have been available to Pappus, then we can start to get an idea of how Pappus is intervening on his sources, what he is including and what he is leaving out. Eratosthenes was, to some extent, "the" name linked with the problem of the two mean proportionals:[38] it is therefore quite expedient for Pappus to suggest parallels between the earlier contribution and his own, and he modifies or selectively reports the other proof in such a way as to create greater continuity with his own demonstration. Three features are given special prominence in Eratosthenes' solution as we find it in the *Collection*: the instrument, the formulation of the problem as cube-duplication and the corollary about finding not just the double but any multiple of a cube. Precisely those three features loom large in Pappus' own construction, which uses an instrument, is presented as a cube-duplication and specifies that it can be used for any other ratio between solid bodies.

Pappus' report of Eratosthenes is concluded thus: "and from this [it is] clear that it is impossible that the question be solved by means of planes."[39] Hultsch dismisses these lines as an interpolation, but no palaeographic reasons are invoked, so he must have done so on grounds of sense – but Pappus' remark makes perfectly good sense. The brief anthology of solutions comes immediately after his classification of problems into planar, solid and linear; therefore this is a strategic position for him to follow up his previous remark (the third objection to the anonymous geometer) that the problem of the two mean proportionals has long been established as solid. With a survey of the four solutions "held up for examination," the reader can satisfy himself that none of them employs planar methods, so that the solid nature of the problem will be confirmed.[40]

---

[38] Even if the report in Eutocius is a forgery, this would only be evidence for the fact that Eratosthenes was renowned enough to be forged.

[39] 58.21–22: Καὶ ἐκ τούτου φανερὸν ὅτι ἀδύνατόν ἐστι τὸ προκείμενον διὰ τῶν ἐπιπέδων λύεσθαι.

[40] 56.8–9: ἐκθησόμεθα οὖν τέσσαρας αὐτοῦ κατασκευάς.

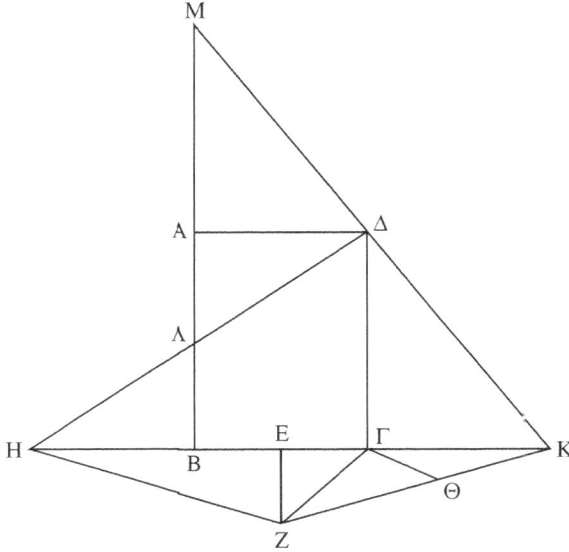

Fig. 4.3. It is required to find two mean proportionals between ΓΔ and ΔA. After completing the parallelogram ABΓΔ, one takes Λ and E as mid-points of AB and BΓ, respectively; prolongs ΔΛ until H and draws EZ perpendicular to BΓ. The point Z is such that ΓZ is equal to ΛΛ, and such that, having joined Z and H, having drawn ΓΘ parallel to it and having prolonged the BΓ until K, so that the angle KΓΘ is given, one draws from the point Z a line ZΘK such that ΘK = ΛΛ = ΓZ. The line in question is a *neusis* and can be determined applying the properties of the cochloid. (In Heath's words (1921), I 236, finding a neusis means "to insert a straight Lne ... of given length ... between [two lines] in such a way that [it] *verges* [in Greek, 'to verge' is νεύειν] towards [a given point]." The procedure was known possibly at the time of Hippocrates of Chios and at least by Archimedes' time, because he mentions it in e.g. *Spirales* 24.2; 28.13.) Having then joined K and Δ and having prolonged KΔ until M, ΓK and MA result to be two mean proportionals between ΔΓ and AΔ.

Nicomedes

Nicomedes' proof is the only instance of a "linear" solution to the problem, i.e. a solution achieved by means of curves such as the spiral, the quadratrix or, in this case, the cochloid. Book 4 of the *Collection* contains an extensive treatment of

4 ALTARS AND STRANGE CURVES

linear curves, which includes Nicomedes' construction, this time preceded by a detailed description of how the cochloid originates. Eutocius' report for its part contains a thorough description of the properties, and instructions for the actual drawing, of the *conchoid*, as it is called there.[41]

The two versions in book 3 and 4 and the one in Eutocius are strikingly similar, yet Eutocius claims to be quoting from Nicomedes' book, while in book 4 Pappus warns us that the demonstration has been "adapted" by him to the construction carried out by Nicomedes. Again in book 4, Pappus refers to another work on the cochloid written by himself, in which the curve is applied to the trisection of the angle.[42] Moreover, Eutocius quotes some introductory remarks where Nicomedes criticizes Eratosthenes' solution, whereas Pappus lists the two proofs one after the other with no mention of any disagreement between their authors.

Analogously to what perhaps happened with Eratosthenes' criticism of previous solutions, it may well be that Pappus decided to leave out Nicomedes' criticism of Eratosthenes because his own concerns were elsewhere: not to provide a history or a comprehensive account of the problem, but to present the three solutions as coexisting without conflict, i.e. as a uniform mathematical tradition, all of a piece with his own practice.[43] But how are the textual coincidences between

[41] In the *Collection*, book 3, at 58.23–62 and book 4 at 246.20–250.25; in Eutocius, *In Sph. Cyl.* 98.1 ff. Knorr (1986a), 226 has Nicomedes' solution derive from the Archimedean-Heronian solution via a lost source. As for the difference of terms between Pappus and Eutocius, although the two authors are definitely talking about the same thing, there has been some discussion as to whether there was a conchoid conceived as something distinct from the cochloid. It is possible that the confusion arose from a scribal error, the groups χλ and γχ being fairly similar in shape, or maybe there just were two similar words for the same thing. Cf. Böker (1962); Knorr (1986a), 222n48.

[42] 246.21–23 (ὁ μὲν Νικομήδης τὴν κατασκευὴν ἐξέθετο μόνον, ἡμεῖς δὲ καὶ τὴν ἀπό-δειξιν ἐφηρμόσαμεν τῇ κατασκευῇ) and 250.26–32, respectively. Cf. also 56.8 and Proclus, *In Eucl.* 272.3–7, saying that Nicomedes "made use of conchoids ... and thus succeeded in trisecting the rectilinear angle generally." See also Seidenberg (1966).

[43] Cf. a different attitude (motivated by the different tenor of the account) in book 6, 556.6 ff., where Pappus reports Aristarchus' results and compares them to Hipparchus' and Ptolemy's, which are not only ἀσυμφώνους with respect to Aristarchus, but also with respect to each other.

the *Collection* and Eutocius to be explained, if the demonstration in Pappus is his own adaptation? One possibility is that Pappus has after all copied the demonstration from Nicomedes and that his own contribution is less radical than it is made to sound; another is that Eutocius has used more than one source, and, in particular, that maybe he used Pappus specifically for the demonstration of the cube-duplication, while the construction of the curve was taken from Nicomedes' own account.

While the version of Nicomedes' solution in book 4 of the *Collection* specifies that the problem is one of cube-duplication, in book 3 there is no trace of corollaries or of the alternative formulation of the problem.[44] That is, the main feature of the construction seems to be simply that it uses a linear curve.

Hero

Hero of Alexandria is mentioned or quoted literally on several occasions in the *Collection*.[45] Both his *Belopoietics* and his *Mechanics* contain solutions to the duplication of the cube, and indeed the heading to the version in Eutocius runs thus: "As (in) Hero, in (the) *Mechanical institutions* and the *Belopoietics*."[46] Pappus' report may have been taken from the *Mechanics*, which is explicitly mentioned in the *Collection*.

The solution is carried out with the help of a moving ruler, which is used to insert a neusis so that the two mean proportionals are found.

Unlike Eratosthenes' or Nicomedes' solutions, it is possible in this case to compare Pappus' and Eutocius' reports with the original – in fact, with two versions of the original. Whereas all the accounts agree on the whole on demonstrative details, it is in prefatory or side-remarks that some differences emerge.

---

[44] 246.18–19.
[45] E.g. 1060.4–10; 1066.31–1068.3; 1070.10–1072.29.
[46] Hero, *Belop.* 114–115; *Mech.* 1.11. Knorr (1986a), 188 ff. and 225 hypothesizes an Archimedean origin for Hero's solution. In Eutocius, *In Sph. Cyl.* 58.15 ff.; Pappus at 62.14–64.18.

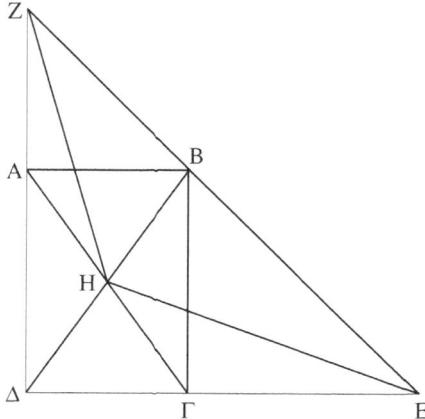

Fig. 4.4. It is asked to find two mean proportionals between AB and BΓ. After having completed the parallelogram ABΓΔ and prolonged the sides until Z and E, one posits a ruler at the point B and moves it until it is in a position such that EH = HZ. AZ and ΓE result to be the two mean proportionals.

Eutocius only reports the proof. The solution in the *Belopoietics* is introduced by Hero, as one would expect, with an illustration of how it can be applied to modify the dimensions of a catapult. It states the possibility both of finding not only the double, but any multiple of a given cube, and of reducing the cube to other solid figures. As for the version in the *Mechanics*, Hero points out the significance of the problem for practical use, then says that it has been carried out "with the help of an instrument, in which case we need no solid figures, and have employed the easiest method for it." The solid figures in question here are (I think) cubes – the context is how to modify the size of two- and three-dimensional figures while respecting their proportions. Hero never spells out the equivalence of the two formulations of the problem, but assures the reader that knowledge of mean proportionals (earlier on in the text he has talked about finding one mean proportional) will avoid dealing with the solid bodies themselves when one wants to alter their dimensions. The *Mechanics* in fact also contains a full proof (rather than just a statement) of the corollary according to which the duplication of the cube can be

applied to proportional modifications of other solid figures.[47]
Pappus quotes it, although the passage raised Hultsch's sus-
picions ("ab aliena manu addita esse apparet").[48] Pappus also
declares that Hero's solution is the most appropriate for
practical applications, first quoting the author himself and
then again remarking that the construction is particularly
handy for architectural purposes. Besides, according to Pappus, Hero himself justifies his choice of a mechanical instrument on the grounds that the problem is solid: "because this
problem is solid, as Hero, too, says."[49]

Yet again, Pappus frames a previous solution in such a way
as to emphasize his favourite issues. Hero's proof again employs an instrument, and its affinity with practice and use in
architecture are brought to the fore. At the same time, the remarks about the fact that the problem is solid, whether they
are originally Heronian or not (none of our extant works by
Hero contains anything in that direction), give Pappus the
opportunity to follow up his preoccupation with labelling
the problem and drawing the consequent methodological
conclusions. By reminding the reader that Hero was indeed
aware of the nature of the problem, and agreed with him,
Pappus is actively constructing a tradition in support of his
classification.

4.3 Stepping out

Pappus' own solution is reported three times in the *Collection*,
as well as once in Eutocius' anthology.[50] The version in book

---

[47] Hero, *Mech.* 1.14 ff. He describes an instrument with the help of which one can
find a plane figure similar to a given one at 1.15 and another instrument for solid
figures at 1.18.

[48] 56.14–17.

[49] 56.11–13 (μάλιστα πρὸς τὰς χειρουργίας ἁρμόζουσαν τοῖς ἀρχιτεκτονεῖν βουλομέ-
νοις); 62.17–18 (ἐκθησόμεθα δέ φησιν τῶν δείξεων τὴν μάλιστα πρὸς τὴν χειρ-
ουργίαν εὔθετον); 62.14–17 (πῶς ἐστιν δυνατὸν δύο δοθεισῶν εὐθειῶν δύο μέσας
ἀνάλογον λαβεῖν ὀργανικῶς, δείξομεν, ἐπειδήπερ ἐστὶν τὸ πρόβλημα τοῦτο, καθά
φησιν καὶ ὁ Ἥρων, στερεόν), respectively. Being apt πρὸς τὰς χειρουργίας is a
claim Pappus makes for instrumental solutions in general; cf. 54.29 and 1070.11.

[50] In book 3, at 64.19–68.16 and 164.1–176.8, and in book 8, at 1070.7–1072.29. In
Eutocius, *In Sph. Cyl.* 70.6–74.31: Ὡς Πάππος ἐν Μηχανικαῖς εἰσαγωγαῖς.

8 has already been mentioned in the previous chapter. The version transmitted by Eutocius indicates its source thus: "As Pappus in the mechanical introductions", which is likely to be either book 8 or a version thereof. That book 8 had a life of its own is testified by the existence of an Arabic translation of it, with the title *Introduction to the science of mechanics*, which could pre-date the earliest extant Greek manuscript itself (*Vat. gr.* 218, tenth century).[51] The second occurrence of the proof in the *Collection*, at the end of book 3, is a long appendix which diverges from the first version and contains several major inaccuracies. Alexander Jones believes that it is a scattered note of Pappus', put together with the rest by some later editor. The authenticity of the passage has universally been questioned, I think on good grounds – apart from the oddity of collocation and the differences in style, it can be found in only one manuscript, and not the most reliable one.[52]

I will thus concentrate on Pappus' first version in book 3, which constitutes the climax of the whole account on the duplication of the cube. The solution is also one of the few cases in which Pappus explicitly introduces a proof as "found by ourselves."[53] We should be careful, however, not to apply our notion of originality to the context of Pappus' work. Even announcing a solution "found by ourselves" does not rule out the possibility that it may have been based on someone else's results. The personal touch may have been a slight modification: an emphasis on particular features, or saying explicitly what instrument was being employed, when this was not specified in the demonstration. In other words, a "discovery" could have been a different positioning of the ruler that avoided a superfluous step, or arranging well-known elements

[51] Cf. Jones (1986a), 9 for the possible versions of book 8; Jackson (1972) and (1980) for Arabic translations of Pappus.
[52] Jones (1986), 25. Cf. Ver Eecke (1933a), XXIV. The manuscript in question is Wolfenbüttel *Gud. gr.* 7.
[53] 56.13: τὴν ὑφ' ἡμῶν ἀνευρημένην. Cf. also the beginning of the proof itself, 64.19–20: εὑρίσκεται ... καθ' ἡμᾶς. The first person plural pronoun is used by Pappus throughout the text (30.19; 32.1; 34.12 and *passim*). Cf. Heath (1921), I 266; van der Eijk (1997).

in such a way as to point out a significant characteristic or an overlooked possibility of application.

As it turns out, Pappus' originality is in question. At the end of his report, Eutocius suggests that Pappus' proof is "the same" as Diocles', except that Diocles uses a point-wise procedure and Pappus a moving ruler.[54] Wilbur Knorr endorsed Eutocius' opinion to the extent of postulating a Dioclean source, now lost, for Pappus.[55]

The fact that Eutocius indicates a link between Pappus and Diocles should not be taken, I think, as historically reliable. Both Pappus' and Diocles' construction are carried out within a circle – but then both Pappus' and Hero's construction use a ruler and both Pappus' and Eratosthenes' construction contain a certain corollary. Eutocius indicates links between quite a few of his sources: the solutions by Hero, Philo and Apollonius, and Diocles, Pappus and Sporus (respectively) are declared to be "the same," even though a superficial analysis is enough to establish that they are not – they share features such as a triangle in the first triplet and a circle in the second. It is also rather interesting that, in the case of the first triplet, Hero's solution is illustrated first, then Philo's, whose construction is perceived as ἡ αὐτή, finally Apollonius', which Eutocius does not even bother to report in detail, because the other two are so similar, he says, that they will suffice.[56] Now, Eutocius was very well aware that Philo could hardly have drawn on Hero. Clearly, the associations the commentator proposes were meant as purely descriptive, as a way of arranging a long list of solutions – not as a suggestion of actual

---

[54] In fact, Diocles' point-wise procedure (διὰ συνεχῶν σημείων) also requires the application of a curved ruler.

[55] Knorr (1982a) and (1989), 11 ff., 81 ff., postulated lost sources for the solutions of Hero, Philo, Apollonius – in short, for a good number of the authors in Eutocius' anthology. The assumption of irretrievable lost sources seems to me to be the equivalent of *ad hoc*, unfalsifiable hypotheses in science (to put it in old-fashioned Popperian terms): it makes one's claims, whether true or not impossible to disprove by evidence. Of the same opinion as Knorr are Böker (1961) and Étienne & Roels (1986). Cf. the (in my view) more balanced opinion in Ver Eecke (1933a), 47n2: "[Eutocius] montre en quoi [la construction de Pappus] ressemble à celle de Dioclès *et en quoi elle en diffère*" (my italics) and (1956), 222.

[56] Eutocius, *In Sph. Cyl.* 62.26 ff.; 66.4–6.

links between them. In other words, Eutocius is not implying that the authors he groups together were in fact depending on each other; he just points out similarities between them, sometimes omitting a proof he judges "the same" as another, sometimes comparing two demonstrations in a sort of mathematical exercise for himself and the reader, as when he compares Diocles' point-wise procedure with Pappus' mechanical procedure; both are aimed at finding a certain point which Eutocius takes to have the same function in both constructions. The only actual connection between Pappus and Diocles, apart from vague similarities in the proof, is the mention in the *Collection* of a curve, the cissoid, which later became associated with the name of Diocles. Nowhere is Diocles explicitly mentioned in Pappus' work – indeed, Eutocius is the earliest author to notify us of the existence of Diocles – but maybe this silence is deliberate.[57] In sum, the case of Pappus' dependence on Diocles remains, in my opinion, undecided – one can no more prove that there is dependence than that there is none.

A rather more likely candidate as source for Pappus' solution could be Sporus, who at least is favourably mentioned by Pappus in book 4.[58] Their two constructions are indeed quite similar, yet the claim that Pappus' solution is his own can still be sustained if we regard some differences from Sporus' proof (as we have it in Eutocius) as changes that make Pappus' contribution different or new. Sporus' construction relies on finding a neusis, but does not specify how the neusis is to be found; Pappus explains that the instrument employed is a moving ruler and describes its use. The point is hardly trivial, since Pappus' proof, as we will see, underlines that the neusis is to be inserted "instrumentally." Again, Sporus' proof makes

---

[57] The cissoid is mentioned at 54.21 and 270.27–28. Ver Eecke (1933a), 39n3, says: "les cissoïdes, dont Pappus attribue la découverte à Dioclès" – but Pappus in fact never does. In fact, according to Toomer (1976), 24, it is "in the highest degree unlikely that the name 'cissoid' was ever applied in antiquity to Diocles' curve" – the connection between the two is not found before the seventeenth century. See also Knorr (1982a), n48.

[58] 252.26–256.1. Heath (1921), I 266, surmises a pupil-teacher link between Pappus and Sporus, but there is no conclusive evidence for it; see also Szabo (1975).

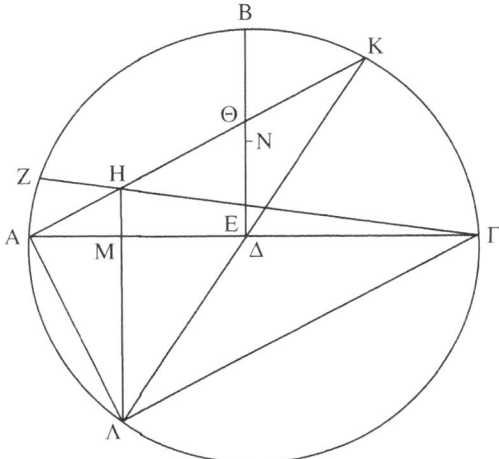

Fig. 4.5. Given the ratio BΔ:ΔE, with a moving ruler we take HΘ = ΘK. It is claimed that the cube on BΔ has to the cube on ΔΘ the ratio BΔ:ΔE. We posit BΔ:ΔΘ = ΔΘ:ΔN (unproblematic because BΔ and ΔΘ are given), and the lines ΔΘ and ΔN will be mean proportionals between BΔ and ΔE.

no mention of the formulation of the problem as a cube-duplication or of any corollaries, whereas Pappus introduces the problem as one of duplication of the cube, and actually mentions the two mean proportionals only at the end of the proposition, assuming that the reader is acquainted with the interchangeability of the two formulations.[59] Moreover, he declares that his proof enables one to find not only the double, but any multiple of a given cube.

Pappus' proof, both here and in book 8, is characterized from the outset as mechanical: he points out that it is carried out "by means of the abovesaid instrument," i e. a moving ruler (κανόνιον), fitted with pivots which enable it to rotate. Its use, Pappus reassures us, is easy – one just has to keep

---

[59] The relevance of choosing one formulation rather than the other is not entirely clear to me. My guess is that an emphasis on the problem as being one of cube-duplication brought out its more mechanical connotations by reminding the reader of its architectural origin.

4 ALTARS AND STRANGE CURVES

trying.[60] For the rest, the solution is carried out by means of the properties of proportions and of triangles inscribed in a circle.

The construction needs only one datum at the outset: the line BΔ, which is the radius of the circle and the first of the two lines between which the two mean proportionals are to be found. The second line, ΔE, is to be given in relation to the first one, and so are the points set as limits for the range of the ruler. Consequently AK and all the other data of the problem are functions of two given lines, ΔE and BΔ. The final result immediately gives us the first of the two mean proportionals, ΔΘ, which is the only datum we actually need for increasing or decreasing cubes or other solids. The results can be visually summarized, as it were, because all the four main elements of the problem lie on the line BΔ, i.e. the same line we started from.

A strong interest on Pappus' part in the practical side of the problem is certainly evident from its placing in book 8, and is in line with the tradition, as exemplified by Eratosthenes, who introduced his solution with a long catalogue of possible ways to use the mesolabe (to enlarge altars, temples and catapults; to measure out dry or liquid goods). As we have mentioned, Eratosthenes flaunted his mathematical expertise by relating the clumsy attempts on the part of previous authors, and was criticized in his turn by Nicomedes, who said that the mesolabe lacked mathematical sense and was ἀμήχανος (inexpedient, unpractical). Correspondingly, in order to establish his expertise, superior with respect to some contemporary geometers, on a par, at least, with the ancients, Pappus puts forth a proof which is meant to be not only methodologically correct, but also εὐμήχανος. The ruler he uses is the same instrument as Hero's; his solution provides a simple and, on a visual level, immediate relation between data and desiderata; it reaps the same benefits as Eratosthenes in terms of corollaries. The location of the proof itself at the end of his anthology is meant

---

[60] 64.19–20 and 1070.14–15: διὰ τοῦ ὑποκειμένου ὀργάνου; 66.12–14: τοῦτο γὰρ πειράζοντες αἰεὶ καὶ μετάγοντες τὸ κανόνιον ῥᾳδίως ποιήσομεν.

to underline the fact that it collects the significant characteristics of the former versions into a new arrangement.

4.4 Facing the alien

The ancient geometers, wanting to divide a given rectilinear angle into three equal parts, had difficulties for this reason. We say that there are three kinds of problems in geometry, and some of them are called planar, some solid, some linear. Those which can be solved by means of straight line and circumference are justly called planar, because the lines by means of which these problems are discovered have their origin in a plane. Those problems whose solution is found applying one or more of the conic sections are called solid, because for their construction it is necessary to employ surfaces of solid figures (I mean it is necessary to employ conic surfaces). A third kind of problems remains, the so-called linear; other lines in fact, apart from the ones we mentioned, are applied to the construction [of a problem]; their origin is more variegated and more constrained and they are generated from more disorderly surfaces and from interwoven motions. Those are the lines found in the so-called loci with respect to a surface, and others more variegated than these and many in number ... having many wondrous properties. More recent geometers have deemed some of these worthy of an extended treatment ... Other lines of this kind are spirals, quadratrices, cochloids, cissoids. It seems to the geometers that it is no small mistake when [the construction of] a planar problem is discovered by someone by means of conics or of linear [curves], and in general when it is solved by means of a kind not its own, as is the case in the fifth book of Apollonius' conics with the problem about the parabola and in Archimedes' spirals about the solid neusis applied to the circle.[61]

[61] 270.1–272.3: Τὴν δοθεῖσαν γωνίαν εὐθύγραμμον εἰς τρία ἴσα τεμεῖν οἱ παλαιοὶ γεωμέτραι θελήσαντες ἠπόρησαν δι᾿ αἰτίαν τοιαύτην. τρία γένη φαμὲν εἶναι τῶν ἐν γεωμετρίᾳ προβλημάτων, καὶ τὰ μὲν αὐτῶν ἐπίπεδα καλεῖσθαι, τα δὲ στερεά, τὰ δὲ γραμμικά. τὰ μὲν οὖν δι᾿ εὐθείας καὶ κύκλου περιφερείας δυνάμενα λύεσθαι λέγοιτ᾿ ἂν εἰκότως ἐπίπεδα· καὶ γὰρ αἱ γραμμαὶ δι᾿ ὧν εὑρίσκεται τὰ τοιαῦτα προβλήματα τὴν γένεσιν ἔχουσιν ἐν ἐπιπέδῳ. ὅσα δὲ λύεται προβλήματα παρςλαμβανομένης εἰς τὴν εὕρεσιν μιᾶς τῶν τοῦ κώνου τομῶν ἢ καὶ πλειόνων, στερεὰ ταῦτα κέκληται· πρὸς γὰρ τὴν κατασκευὴν χρήσασθαι στερεῶν σχημάτων ἐπιφανείαις, λέγω δὲ ταῖς κωνικαῖς, ἀναγκαῖον. τρίτον δέ τι προβλημάτων ὑπολείπεται γένος τὸ καλούμενον γραμμικόν· γραμμαὶ γὰρ ἕτεραι παρὰ τὰς εἰρημένας εἰς τὴν κατασκευὴν λαμβάνονται ποικιλωτέραν ἔχουσαι τὴν γένεσιν καὶ βεβιασμένην μᾶλλον, ἐξ ἀτακτοτέρων ἐπιφανειῶν καὶ κινήσεων ἐπιπεπλεγμένων γεννώμεναι. τοιαῦται δὲ εἰσιν αἵ τε ἐν τοῖς πρὸς ἐπιφανείαις καλουμένοις τόποις εὑρισκόμεναι γραμμαὶ ἕτεραι τε τούτων ποικιλώτεραι καὶ πολλαὶ τὸ πλῆθος ... πολλὰ καὶ θαυμαστὰ συμπτώματα περὶ αὐτὰς ἔχουσαι. καὶ τινες αὐτῶν ὑπὸ τῶν νεωτέρων ἠξιώθησαν λόγου πλείονος ... τοῦ δὲ αὐτοῦ γένους ἕτεραι ἕλικές εἰσιν τετραγωνίζουσαί τε καὶ κοχλοειδεῖς καὶ κισσοειδεῖς. δοκεῖ δὲ πως ἁμάρτημα τὸ τοιοῦτον οὐ μικρὸν εἶναι τοῖς γεωμέτροις, ὅταν ἐπίπεδον

The two propositions at issue have been identified with prop. 51 of Apollonius' *Conics* (on lines normal to a parabola) and prop. 18 of Archimedes' *Spirals*.[62] Pappus is our only source for such a detailed classification of problems, and especially for a classification with prescriptive value, but there is evidence for earlier, analogous distinctions between *loci*.[63] A line of tradition which has left traces in Eutocius and Proclus indicates that *loci* were divided into planar and solid (in Proclus), or into planar, solid and "*loci* with respect to a surface" (in Eutocius). Eutocius grouped together conic sections and "many others" in his definition of solid *loci*; Proclus did the same with conic sections and the cylindrical spiral.[64] In Knorr's opinion, in early classifications of *loci* "all cases generated from solids [were conflated], rather than differentiating the 'solid' from the 'linear' type,"[65] as indeed indicated by the existence of some linear solutions to solid problems (Diocles and Nicomedes duplicated the cube by means of cissoid and cochloid, respectively, between the third and second century BC). In book 4, Pappus points out that the trisection of an angle is a solid problem, but the division of a given angle or circumference according to any given ratio is linear, as "proved by more recent authors."[66] Again, in a passage from book 7 of the *Collection*, where Pappus is drawing on Apollonius' books on planar *loci*, linear *loci* are first put together with planar and solid ones, being all διεξοδικοὶ σημεῖων, literally "path *loci*," i.e. generated from the movement of a

---

πρόβλημα διὰ τῶν κωνικῶν ἢ τῶν γραμμικῶν ὑπό τινος εὑρίσκηται, καὶ τὸ σύνο-λον ὅταν ἐξ ἀνοικείου λύηται γένους, οἷόν ἐστιν τὸ ἐν τῷ πέμπτῳ τῶν Ἀπολλωνίου κωνικῶν ἐπὶ τῆς παραβολῆς πρόβλημα καὶ ἡ ἐν τῷ περὶ τῆς ἕλικος ὑπὸ Ἀρχιμήδους λαμβανομένη στερεοῦ νεῦσις ἐπὶ κύκλον. This passage is in book 4 and comes after the discussion of the quadratrix. The same passage in book 3 at 54.7 ff., coming after the criticism of the anonymous geometer's solution to the problem of the two mean proportionals. Cf. Tannery (1883); Knorr (1978b) and (1986a), 345.
62 Knorr (1986a), 319–320.
63 According to Jones (1986a), II 395, *loci* are "definable geometrical object[s] on which any point or line satisfying the conditions will be found." He also adds: "the 'solution' or 'demonstration' of a locus is the construction of that object [point or line], and proof that it is indeed the locus."
64 Eutocius, *In Apol. Con.* I 184.21–28; Proclus, *In Eucl.* 394.1–395.12.
65 Knorr (1986a), 342. See also Tannery (1883–4).
66 284.21–25.

point,[67] but then Pappus remarks that linear *loci* are demonstrated on the basis of *loci* with respect to surfaces, which in their turn are path *loci* with respect to lines, but ἀναστροφικοί (domain *loci*, literally *loci* of revolution) with respect to points. The characterization of *loci* with respect to surfaces is further complicated if we take on board Eutocius' definition, according to which they were identified by their specific characteristic.[68] Returning to book 7 in the *Collection*, a few lines below, we are told that the people "who came after the ancients" added more *loci* to the planar ones, which were distinguished and systematized in early compilations of elements. Planar *loci*, Pappus continues, are about straight lines and circles, solid ones about conic sections, linear ones are neither of the above.

It would thus seem that the items belonging to the linear category were known at quite an early stage, but their definition *as* linear came later. It may also be that the threefold distinction was first applied to *loci* and only transferred to problems at a later stage, possibly at the hands of Pappus himself. In any case, earlier classifications seem to have had more of a descriptive or taxonomic value than a prescriptive one. The prescriptive value of the classification is emphasized by its collocation in book 3, where the anonymous planar solution to the solid problem of the two mean proportionals is sharply criticized.

Linear curves – spiral, quadratrix, cochloid – are the main topic of the last part of book 4. For each of them, Pappus systematically describes the movements from which the curve originates, its main, identifying property (σύμπτωμα) and its applicability to one or more geometrical problems. The definition of linear curves as a group is nevertheless rather tentative; also, several of the expressions used by Pappus to define the origin of linear curves (ποικίλος, βεβιασμένη, ἄτακτος) had negative undertones in common Greek usage. These terms were usually associated with the irregular, the unnatural or

---

[67] 660.17–664.23.
[68] Eutocius, *In Apol. Con.* 184.21–28: εἰσὶ δὲ καὶ ἄλλοι ἐπωνυμίαν ἔχουσι ἀπὸ τῆς περὶ αὐτοὺς ἰδιότητος.

the anomalous and denoted "strange" or not easily definable entities.[69] In other words, the curves Pappus is dealing with were a bit of an oddity. He tries to give them the status of re-spectable mathematical objects, combining displays of their versatility as problem-solving tools with appeals to the au-thority of mathematicians old and new; he refers to various authors, including perhaps Euclid, who have studied them, and remarks that "the more recent [geometers] have deemed some of these [curves] worthy of an extended treatment."[70] The features of Pappus' account will emerge more clearly if we compare it with analogous classifications in Geminus (*apud* Proclus) and Hero.

Geminus (first century BC) is explicitly quoted in the *Col-lection* once, in book 8.[71] He was allegedly close to Stoic cir-cles, and indeed some of the criteria and terms he and/or Proclus use for classification are philosophy-laden to a marked degree. For instance, Proclus talks about the πάθος that "affects" a spiral (to denote what elsewhere would be called σύμπτωμα) or indicates the various ways in which ge-ometrical objects can be mixed as μῖξις, κρᾶσις and σύνθεσις.[72] The first distinction Geminus makes is between composite and incomposite (σύνθετον and ἀσύνθετον) lines, then within the incomposite lines he identifies two further pairs of categories: σχῆμα-generating or prolongable *ad infinitum* on the one

---

[69] *LSJ ad voces.* Cf. also Detienne & Vernant (1974).

[70] 270.1–28. The authors mentioned are Demetrius Alexandrinus (20), Philo Ty-naeus (21) and Menelaus (26). Euclid is not named, but Pappus refers (18–19) to the book *Loci with respect to a surface* which is attributed to Euclid in book 7, 636.23–4. The quotation is at 24.25: καί τινες αὐτῶν ὑπὸ τῶν νεωτέρων ἠξιώθησαν λόγου πλείονος.

[71] 1026.6–9. Pappus is quoting from *On the order of mathematics.* Another connec-tion between Geminus and Pappus can be traced: a couple of passages, again in book 8 (1110.14 ff. and 1124.5 ff.), describe cylindrical spirals, which are said to have been investigated by Apollonius. The only other occurrence of cylindrical spirals in connection with Apollonius is in Proclus, *In Eucl.* 104.26–106.9, when the latter reports a criticism of Apollonius' claim that the cylindrical spiral is a simple line. The significant fact is that Proclus' source for the criticism is Geminus. The coincidence of topics indicates that *The order of mathematics* known to Pap-pus may have been the same text that deals with the classification of lines reported by Proclus. The only book title by Geminus quoted by Proclus is the *Philokalia*, at *In Eucl.* 177.24. On Geminus in general see Tittel (1910); Heath (1921); Dicks (1972).

[72] Proclus, *In Eucl.* 105.7 and 117.25 ff., respectively.

hand, and simple or mixed (ἁπλαῖ and μικταί) on the other.[73] In the first case, the classification cuts across categories that in Pappus are unified: cissoid and conchoid are both in-composite, but the first belongs to the σχῆμα-generating lines, whereas the second is prolongable *ad infinitum*. Parabola and hyperbola (called with early terminology the sections of a right-angled and of an obtuse-angled cone) fall on different sides to the ellipse (defined as "oval figure", θυρεός), because they are respectively prolongable *ad infinitum* and σχῆμα-generating.[74] The classification brings together such strange bedfellows as circumference and cissoid on one side, and conchoid and straight line on the other.

As for the pair simple/mixed, we find it in the context of a philosophical discussion about the nature of simple lines. Proclus starts by citing Plato and Aristotle and ends with Geminus' distinction between simple and merely homogeneous lines, and his consequent criticism of Apollonius' idea that the cylindrical spiral is a simple line.[75] While "simple" was a received category, access to which was traditionally conceded only to a few unproblematic lines, "mixed" allowed for a host of definitions (not necessarily mutually exclusive). According to one of them, "mixed" was used with reference to the motions from which a line originated. Even apparently straightforward objects, such as the diagonal of a square, are thus described in Proclus as the result of two motions, one lengthways and one sideways.[76] Thus, in this latter case "simple" amounts to "compounded by similar motions," whereas "mixed" is "compounded by dissimilar motions." Under this description the ellipse, for instance, is a mixed line – the motions generating it are characterized as taking place non-uniformly (ἀνωμάλως) and not according to nature (μὴ κατὰ φύσιν).[77] Again, within the field of mixed lines some

---

[73] Proclus, *op. cit.* 111.1–113.6 and 176.27–177.23.
[74] Proclus, *op. cit.* 111.6; cf. Ver Eecke (1948), 100n2; Knorr (1982a), (1982b) and (1992).
[75] Proclus, *op. cit.* 105.25 ff.
[76] Proclus, *op. cit.* 106.3 ff.
[77] Proclus, *op. cit.* 106.14 and 18.

are in planes, like the cissoid, some in solids, and of these, some are imagined "according to sections of solids," some "around" the solids.[78] Conic sections belong to the category of "according to sections of solids"; spherical and conical spirals, instead, to the group of "around the solids." As for conchoids, Geminus/Proclus refers rather vaguely to lines of variegated shape ("variegated" is here ποικίλος) that are produced by cutting solids, and the conchoid is one of them.[79] Incidentally, neither Pappus nor Eutocius says that the cochloid/conchoid can be obtained by cutting a solid.

Proclus also declares that Geminus had transmitted the origin of conchoids, of cissoids and of spirics, adding that the students who love knowledge (or mathematics, φιλομαθεῖς) can refer to Geminus' book for the proofs, since he (Proclus) considers it "superfluous in the present work ... to make a precise enquiry into each of them.... Nowhere does he ["our geometer", i.e. Euclid] mention mixed lines. And yet he knows mixed angles ... mixed plane figures ... and mixed solids."[80]

In sum, in the commentary on Euclid the strange lines which are the object of Pappus' enquiry in book 4 of the *Collection* defy unifying definitions and are marked out by their irregular or unnatural origin. Conchoids and spirals are not sharply separated from conic sections and, given that what is left out from Proclus' report is a description of the origin of these curves, there seems to be no indication that their use in problem-solving had been made the object of systematic enquiry.

Hero's *Definitions*, which also contains distinctions between lines, is a composite piece of work, more of a compilation than a systematic treatment, and its question-and-answer structure suggests that it was meant as a handbook or a

---

[78] Proclus, *op. cit.* 111.15–17.
[79] Proclus, *op. cit.* 112.15.
[80] Proclus, *op. cit.* 113.3–19. Proclus also declines to discuss mixed lines with reference to trisection or division by any ratio of angles and arcs, *op. cit.* 272.12–14: "The thoughts of these men are difficult for a beginner to follow, and so we pass them by here."

schooltext.[81] Given his knowledge of other works by Hero, there is the possibility that Pappus was acquainted with its general contents. The book sets out two main types of lines: straight lines and all the rest;[82] the "rest" are in turn divided into circular, helicoidal and curved lines. This latter category should include conic sections, conchoids and, one infers, many others, since their number is infinite.[83] Later in the text, another distinction is applied to lines in solid figures, which divide into simple and mixed. Conics and spirics are assigned to the second category. The author further specifies: "and these are the ordered ones, whereas the number of the unordered ones is infinite, as is the number of the composite ones."[84] One can see the same terminology recurring in Geminus (Proclus) and Hero. They seem to be at a loss for more precise vocabulary to circumscribe a set of objects with a still unspecified number of ragged ends – to chart a region whose borders fade into indistinctness.

The existence of other accounts which share some terminology with Pappus indicates that his classification did not spring up out of nowhere, but is to be inscribed in an already existing tradition, although I do not think that Pappus' account of linear curves owes a substantial debt to any source in particular. There seems to be no received notion of what spirals or conchoids were *as a group*, characterized by something other than their otherness. Pappus himself sets no definite conditions for a curve to be linear: he just lists the most famous ones and adds "and others ... in great number." Linear curves are identified on the strength of their problem-solving performance, on the basis, so to speak, of their function; their essence (being generated in a varied way, etc.) is left vague, but their place in the classification of problems and related legitimate procedures is firmly established. The three curves which are chosen by Pappus for his account in book 4

[81] Knorr (1993).
[82] Hero, *Definitiones* 16.18–20, def. 3.
[83] Hero, *op. cit.* 18.13–14, def. 6.
[84] Hero, *op. cit.* 52.3 ff., def. 75. Quotation at 52.8–9: καὶ αὗται μὲν τεταγμέναι εἰσίν, τῶν δὲ ἀτάκτων πλῆθος ἄπειρόν ἐστιν ὡς καὶ τῶν συνθέτων.

were, arguably and according to other ancient testimonies, the most fruitful ones as far as results in problem-solving were concerned.[85] Also, at least in the case of the Archimedean spiral, they had the appeal of a great name attached to them. We shall examine them one by one.

Spiral

The spiral (ἕλιξ) was probably the most famous of the three curves, being the subject of a work by Archimedes. Just before defining the curve, Pappus acknowledges the work of his predecessors: the geometer Conon of Samos, who proposed the main theorem about the spiral, and Archimedes who proved it using a "wondrous conception."[86]

Pappus' treatment has been the topic of an influential paper by Knorr, who claims that its divergence from Archimedes' work (different phrasing at crucial points, six propositions in Pappus instead of the twenty in Archimedes) is due to the fact that he is drawing on an early version of the *Spirals*, a version which Archimedes himself subsequently modified in response to criticisms similar to those voiced by Pappus in the passage we have quoted.[87] My opinion is that postulating a lost source is not absolutely necessary – the contrasting features of the two treatments could well be due to the fact that they belong to different times and authors and have different aims. The lapse of time between the third century BC and the fourth century AD had seen interest shift from the study of spirals in two-dimensional space, with an emphasis on expressing curvilinear areas, to that of spirals in three-dimensional space, their use in loci with respect to surfaces, and, most of all, their potential as problem-solving tools (they had been successfully applied to a number of problems, including the general divi-

---

[85] Proclus, *In Eucl.* 272.3 ff.; Simplicius, *In Arist. Phys.* 60.7–16.

[86] 234.1–4: Τὸ ... θεώρημα προὔτεινε μὲν Κόνων ὁ Σάμιος γεωμέτρης, ἀπέδειξεν δὲ Ἀρχιμήδης θαυμαστῇ τινι χρησάμενος ἐπιβολῇ. Conon (if indeed it is the same person) is mentioned in the prefaces of several works by Archimedes: *Sphere and Cylinder I* and *II, Spirals, Quadrature of the Parabola*.

[87] Knorr (1978a) and see also Vitrac (1992).

sion of an angle), or even as the geometrical structures of mechanical devices.

The tradition on the spiral which was available to Pappus had certainly retained the priority of Archimedes and his contributions as highly important. Even so, those contributions may have been modified – in the course of the tradition, or consciously by Pappus himself. Given the overall orientation of his treatment of linear curves, one would expect Pappus to be less concerned about faithfully reporting everything Archimedes had said than about making the spiral fit into his own system. This is exactly what he does. Archimedes has a general proposition thus:

> If a straight line is drawn in a plane and, while one of its extremities remains at rest, the other goes round with uniform velocity (ἰσοταχέως) until it goes back to the point from which it started, and at the same time as the line goes round a point moves with uniform velocity along the line from the extremity that remains at rest, the point describes a spiral in the plane.

Pappus, on the other hand, has:

> Let there be a circle whose centre is B and whose radius is BA. Let the line BA be moved so that the point B remains at rest and the point A is carried uniformly (ὁμαλῶς) along the circumference of the circle, while at the same time as this let a point be carried uniformly on it, beginning at the point B towards the point A, and let this point be carried across the line BA in the same time as the point A is carried across the circumference of the circle. So the point moved along the line BA describes in its going around a line, such as BEZA ... this line is called spiral and its main property is this ...[88]

Pappus' description of a curve is always specific (the reference is to a lettered diagram) rather than general as in Archimedes; the curve is presented in terms of given geometric elements which move tracing certain paths, and there is always a specification of the main property (ἀρχικὸν σύμπτωμα).

---

[88] Archimedes, *Spirales* def.1; Pappus at 234.5-18: ῎Εστω κύκλος οὗ κέντρον μὲν τὸ Β, ἡ δὲ ἐκ τοῦ κέντρου ἡ ΒΑ. κεκινήσθω ἡ ΒΑ εὐθεῖα οὕτως ὥστε τὸ μὲν Β μένειν, τὸ δὲ Α ὁμαλῶς φέρεσθαι κατὰ τῆς τοῦ κύκλου περιφερείας, ἅμα δὲ αὐτῇ ἀρξάμενόν τι σημεῖον ἀπὸ τοῦ Β φερέσθω κατ᾽ αὐτῆς ὁμαλῶς ὡς ἐπὶ τὸ Α, καὶ ἐν ἴσῳ χρόνῳ τό τε ἀπὸ τοῦ Β σημεῖον τὴν ΒΑ διερχέσθω καὶ τὸ Α τὴν τοῦ κύκλου περιφέρειαν· γράψει δὴ τὸ κατὰ τὴν ΒΑ κινούμενον σημεῖον ἐν τῇ περιφορᾷ γραμμὴν οἵα ἐστὶν ἡ ΒΕΖΑ ... αὐτὴ δὲ ἡ γραμμὴ ἕλιξ καλεῖται. καὶ τὸ ἀρχικὸν αὐτῆς ἐστι σύμπτωμα τοιοῦτον. I thank R. Netz for his help with the translation of this passage

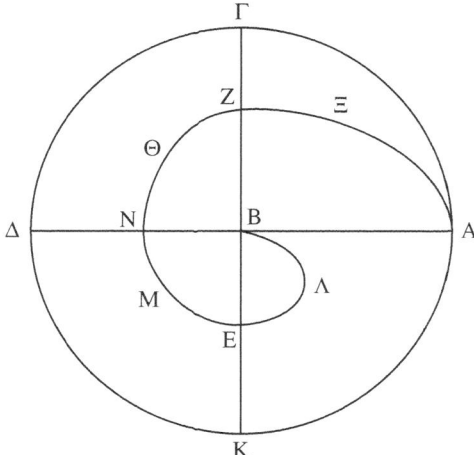

Fig. 4.6. The spiral is obtained when in the circle AΔΓ the line BA moves with its extremity A along the circumference, while at the same time a point on it goes from B to A. If we posit the space between BΛE and the line BE as 1, the areas limited by the spiral and by the various segments of diameters will have to each other the ratio 4:3:2:1.

The theorems in Pappus' account of the spiral correspond, in fact, to the theorems Archimedes picks up for his introductory letter to the *Spirals*. After recalling various results of *Sphere and Cylinder* and *Conoids and Sphaeroids*, the Syracusan geometer cites four of the propositions contained in the book, which are, he warned Dositheus, "problems of a kind which has nothing in common with the ones mentioned above." They correspond to prop. 24, 18, 27 and 28 of our extant text.[89] Pappus' account consists in fact of prop. 24, with a different demonstration; of a corollary of prop. 24; of a theorem whose mathematical import is analogous to that of prop. 28, and, finally, of an equivalent of prop. 27 which amounts to an "arithmetization" of the results Pappus has obtained so far (he establishes not only the ratios of various portions of the area bounded by the spiral to each other in general, but their numerical relations as well).[90]

[89] *Spirales* 2.2 ff., esp. 8.13 ff.   [90] 242.1–12.

The only theorem which was in Archimedes' introduction but is left out in the *Collection* is prop. 18 – the same proposition criticized in the passage we have quoted.

Cochloid

The discovery of the cochloid (κοχλοειδής) is unanimously attributed to Nicomedes, who probably lived in the second half of the third century BC. The most extensive reports about it are found in Pappus and Eutocius, as we have already seen.[91] Apart from calling the curve conchoid (κογχοειδής) and reporting Nicomedes' criticism of Eratosthenes' solution to the problem of the two mean proportionals, Eutocius, unlike Pappus, also provides a very detailed description of the device used to draw the curve (two rulers perpendicular to each other and connected by a pivot so that one of them can rotate) and a full demonstration of two of its properties. Pappus for his part describes the curve, but gives no actual instructions as to its construction; he also points out the two properties reported by Eutocius, but gives a full proof only of the second one, which is related to how the cochloid can be used to find a generic neusis.

As we have seen, there were several possible ways of defining or describing curves: they could be viewed as sections of some oddly shaped solid or as a succession of points, connected with the help of a curved ruler, with an implicit appeal to the principle of continuity. They could be defined exclusively by reference to the mechanical instrument used to draw them, or pictured as intersections of the paths of two movements. Clearly, it was nearly always possible to describe a curve in any of the ways reported above – therefore, the choice of one or the other implied a deliberate emphasis on some characteristics rather than others.

The definition of the conchoid in Eutocius coincides with its

---

[91] The curve is also mentioned in Proclus, *In Eucl.* 272.3 ff. and 356.8 ff.; Simplicius, *In Arist. Cat.* 192.15–25 and *In Arist. Phys.* 60.7–16. See in Proclus what seems an attribution of its discovery to Geminus: *In Eucl.* 113.4 ff. See also Knorr (1986a), 219.

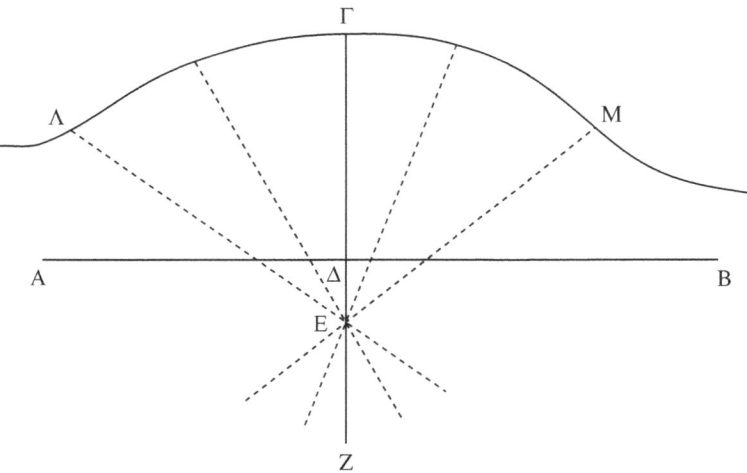

Fig. 4.7. A line AB and another line ΓΔZ perpendicular to it are given. We take a point E on the line ΓΔZ and, while E remains where it is, the line ΓΔEZ moves along the line AΔB so that the point Δ remains along the line AB. The point Γ will describe a curve ΛΓM, which we call cochloid. The line AB is also called axis (κανών). Its main property is that, if one draws a line from the point E towards the curve, the part of this line comprised between AB and the curve is equal to ΓΔ.

instrumental construction – the line comes into existence even as it is drawn, following the instructions to slide one ruler or the other. For all his three curves, Pappus adopts instead a geometrical mode of exposition: the cochloid is defined by motions of points and lines in a way that, if not abstract in itself, at least abstracts from the fact that, in the actual production of the curve, these lines "are" rulers sliding along other rulers. This process does not entail the rejection of mechanical aids – although they do not enter the definition as such, they are indispensable as a means of construction. Pappus does not fail to specify that the cochloid is drawn by means of an instrument.[92] Assuming that the differences between the report in the *Collection* and that in the commentary on Archimedes are the result of Pappus' modification of his

[92] 244.21.

source, then, I take it that the description of the cochloid in terms of the intersection of two paths is meant to create continuity with the other two curves, which are presented in a similar fashion. At the same time, Pappus may be trying to make a connection with authoritative earlier accounts. For instance, his remark that there are also a second, third and fourth cochloid, which "are useful for other theorems," suggests a parallel between the cochloid and the spiral – in Archimedes' treatment there are a second line, a third line, area and circle (all relative to the spiral) and so on, which acquire the same name as the number of revolutions completed by the original point.[93]

Given its σύμπτωμα, the cochloid naturally lends itself to the solution of problems that require the insertion of a neusis, because it enables one to produce segments all of a specified length. Alternative methods employed the intersection of a hyperbola and a parabola or, in some cases, of a hyperbola and a circumference. We can then imagine that the cochloid represented a valid substitute for conic sections: both procedures relied to some extent on the use of instruments, so the difference between the two methods was not a matter of whether or not mechanical means were used (and that was not a problem anyway). Rather, as far as practicality went, the cochloid procedure may have been more convenient because it implied the (standardized) construction of one curve rather than two. We have seen how the cochloid is applied to the duplication of the cube; we are also told that it had been applied to the trisection of the angle (another solid problem) in Pappus' commentary on Diodorus' *Analemma*.[94]

Quadratrix

The quadratrix (τετραγωνίζουσα) owes its name to the problem for which it was apparently devised, the squaring of the circle.[95] The problem had received a great deal of attention from philosophers (Aristotle, to name but one); there were

---

[93] 244.18–20: εἰς ἄλλα θεωρήματα χρησιμεύουσαι. Archimedes at *Spir.* 46.1 ff.
[94] 246.1–3.

very accurate and authoritative arithmetical approximations of the ratio between diameter and circumference, and a formulation of the area of the circle had been given by Archimedes and was standard by Pappus' time, if not long before.[96]

Rather than ascribing the discovery of the quadratrix to anyone in particular, Pappus specifies that the curve had been applied to the quadrature of the circle by Deinostratus, Nicomedes and some other, more recent, authors.[97] The report in the *Collection* need not clash with the one in Proclus, who attributes the demonstration of the σύμπτωμα of the curve to a Hippias, yet refers to the quadratrices "of Hippias and Nicomedes," which, he says, had been applied to the general division of the angle.[98]

As he has done for the cochloid, Pappus gives the description of the quadratrix in terms of the motions which generate it and of its main property. The construction, however, presents some problems. Pappus quotes two objections by Sporus which he terms "reasonable" (εὐλόγως). The first is that, unless the ratio of the velocities of AB and BΓ is already given, or unless the ratio of a radius to a quadrant is already given (Sporus seems to assume that the two are equivalent), the curve cannot be constructed. We are in a context where the quadratrix is being devised in order to square the circle, so obviously the second condition (that the ratio of a radius to a quadrant be given) cannot be satisfied without circularity. As a conclusion, the construction does not work. In Sporus'/ Pappus' words:

How is it possible, if two points start to move from the point B, one along a straight line towards A, the other along a circumference towards Δ, to come

---

[95] Knorr (1986a), 84–5, thinks that the curve was first devised to trisect the angle.

[96] Aristotle, *Analytica Posteriora* 75b41; *Physica* 185a14–17; *Sophistici Elenchi* 171b13–17, and cf. Mueller (1982). Approximations had been carried out by Archimedes, Apollonius (according to Eutocius), Ptolemy, Philo of Gadares (again, according to Eutocius); all quoted in Heath (1921), I 232 ff. The standard formulation in Archimedes, *Measurement of the circle* 1.

[97] 252.1–2.

[98] Proclus, *In Eucl.* 356.11 and 272.7 ff., respectively. Whether Proclus' Hippias is to be identified with the contemporary of Socrates is a matter for debate, cf. Knorr (1986a), 80–1. Pappus has a solution to the general division of the angle in which a quadratrix is employed at 286.1–18.

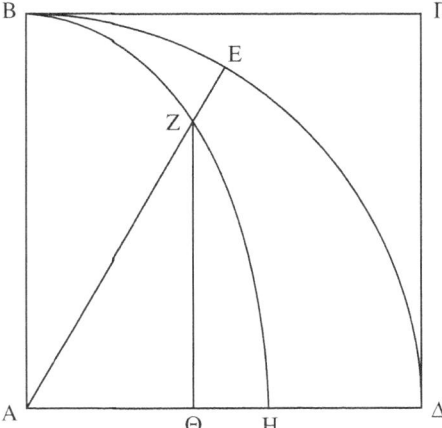

Fig. 4.8. The extremity B of the line AB moves along the circumference BEΔ. At the same time the line BΓ goes down, always parallel to itself, i.e. with B moving along BA. The quadratrix BH results from the two paths intersecting.

back to the same point at the same time, unless the ratio of the line AB to the circumference BEΔ is known before? Indeed it is necessary that the velocities of the movements have this same ratio as well. Since in what way [is it possible] to come back to the same point having indeterminate velocities, except if it happens by some chance? Isn't that absurd?[99]

Sporus' second objection is to the effect that there is no definite way of indicating the limit-point of the intersection between the two paths, *ergo* the quadratrix itself remains undetermined. To quote again:

The extremity of the curve which is used for the squaring of the circle, that is the point by which the line AΔ is cut [*sc.* the point H], is not found.... Even if the lines ΓB and BA, having been drawn, come back to the same point, they will coincide with the line AΔ and will produce no section by intersecting each other. The intersection in fact ends before the coincidence

---

[99] 254.2–10: πῶς γὰρ δυνατόν, δύο σημείων ἀρξαμένων ἀπὸ τοῦ Ε κινεῖσθαι, τὸ μὲν κατ' εὐθείας ἐπὶ τὸ Α, τὸ δὲ κατὰ περιφερείας ἐπὶ τὸ Δ ἐν ἴσῳ χρόνῳ συναποκαταστῆσαι μὴ πρότερον τὸν λόγον τῆς ΑΒ εὐθείας πρὸς τὴν ΒΕΔ περιφέρειαν ἐπιστάμενον; ἐν γὰρ τούτῳ τῷ λόγῳ καὶ τὰ τάχη τῶν κινήσεων ἀνάγκη εἶναι. ἐπεὶ πῶς οἷόν τε συναποκαταστῆναι τάχεσιν ἀκρίτοις χρώμενα, πλὴν εἰ μὴ ἂν κατὰ τύχην ποτὲ συμβῇ; τοῦτο δὲ πῶς οὐκ ἄλογον;

along the line ΑΔ, the same intersection which instead should become the limit of the curve, in which the two lines meet the line ΑΔ. Unless someone says that the curve is imagined prolonged until it meets ΑΔ – this does not follow from the principles set at the beginning, but rather would take the point Η having taken as a premise the ratio of the line to the circumference.[100]

This also shows that there was no standard procedure to draw the quadratrix. Sporus' criticism echoes Pappus' attack on the anonymous construction of the two mean proportionals: the issue is again determinacy; the difficulty is expressed in analogously literary tones (e.g. the use of a couple of rhetorical questions), and once again, after the first objection, the author makes concessions ("even if the lines ΓΒ and ΒΑ, having been drawn, come back to the same point") only to show that the construction is fallacious in any case.

Having thus warned the reader about the difficulties stemming from the construction of the curve, Pappus shows how it would be used to square the circle, and then proposes to analyze the origin of the quadratrix geometrically, since it had previously been presented, he says, in a more mechanical way, i.e. in the way implying problems with velocities which had been criticized by Sporus.[101] An alternative has to be sought, and Pappus provides two propositions which are introduced as using *loci* with respect to surfaces. Both seek to determine the intersection point of the two paths via a point-wise procedure, thus avoiding accusations of indeterminacy, and both

[100] 254.10–22: τὸ πέρας αὐτῆς ᾧ χρῶνται πρὸς τὸν τετραγωνισμὸν τοῦ κύκλου, τουτέστιν καθ' ὃ τέμνει σημεῖον τὴν ΑΔ εὐθεῖαν, οὐχ εὑρίσκεται.... ὁπόταν γὰρ αἱ ΓΒ ΒΑ φερόμεναι συναποκατασταθῶσιν, ἐφαρμόσουσιν τῇ ΑΔ καὶ τομὴν οὐκέτι ποιήσουσιν ἐν ἀλλήλαις· παύεται γὰρ ἡ τομὴ πρὸ τῆς ἐπὶ τὴν ΑΔ ἐφαρμογῆς ἤπερ τομὴ πέρας αὖ ἐγένετο τῆς γραμμῆς. καθ' ὃ τῇ ΑΔ εὐθείᾳ συνέπιπτεν. πλὴν εἰ μὴ λέγοι τις ἐπινοεῖσθαι προσεκβαλλομένην τὴν γραμμήν, ὡς ὑποτιθέμεθα τὰς εὐθείας, ἕως τῆς ΑΔ· τοῦτο δ' οὐχ ἕπεται ταῖς ὑποκειμέναις ἀρχαῖς, ἀλλ' ὡς ἂν ληφθείη τὸ Η σημεῖον προειλημμένου τοῦ τῆς περιφερείας πρὸς τὴν εὐθεῖαν λόγου. Heath (1921), I 229–30 thought that "both Sporus's objections are valid." Cf. *contra* van der Waerden (1954), 192.

[101] 258.20–5: Αὕτη μὲν οὖν ἡ γένεσις τῆς γραμμῆς ἐστιν, ὡς εἴρηται, μηχανικωτέρα, γεωμετρικῶς δὲ διὰ τῶν πρὸς ἐπιφανείαις τόπων ἀναλύεσθαι δύναται τὸν τρόπον τοῦτον. At 254.22–24 (immediately after Sporus' criticism) we had: χωρὶς δὲ τοῦ δοθῆναι τὸν λόγον τοῦτον οὐ χρὴ τῇ τῶν εὑρόντων ἀνδρῶν δόξῃ πιστεύοντας παραδέχεσθαι τὴν γραμμὴν μηχανικωτέραν πως οὖσαν.

employ spirals, respectively a spiral in a cylinder and a spiral in a plane.[102] The quadratrix is also applied to the division of arcs or angles according to any ratio – the only instance we have in the *Collection* of a declaredly linear problem. Pappus solves it in two different ways, first by means of a quadratrix and then by means of a spiral – the linear nature of the problem is thus reaffirmed by showing that it can be solved using two out of the three linear curves Pappus has chosen for examination.[103] The quadratrix is then applied to the construction of an arc which has the same extremities as a given line and has a given ratio to the line itself, and to the connected construction, within a quadrant, of an angle incommensurable to a given one. The variety of problems for which the curve can be employed, some of an apparently digressive nature, reinforces the impression that Pappus, far from emphasizing the role of the quadratrix for squaring the circle, wished to provide a substantial number of examples where the curve was put to a different use.[104]

Conclusion

Let us recapitulate: we have seen how Pappus relates to a classical problem, on which a long tradition of research existed, and to a not-so-classical topic, which had been treated before but was not fully established. Both accounts were meant for people already fairly specialized in mathematics, that is, people who were already acquainted with the issues involved. Pandrosion is a teacher of mathematics; the people to whom the anonymous construction of the two mean proportionals had been sent and the philosopher Hierius himself were clearly able to understand what was going on; the addressee of book 4 (whose first part deals with paradoxical theorems, in a sort of *divertissement* which would have fitted

---

[102] 258.20–262.2 and 262.3–264.2.
[103] 284.21–288.3.
[104] Molland (1976), 27: "It seems clear that Pappus regarded the spiral and the cylindrical helix as having a firmer claim to the status of being geometrical than the quadratrix."

in well with the strangeness of the linear curves in the last part of the account) is supposed to be acquainted with Archimedes' *Spirals*, as we have already observed in chapter two.

Pappus' approach is geared to his subject-matter. In the case of the two mean proportionals, he is keen to stress his role as custodian and successor to the earlier geometers, so that his direct interventions are very explicit: he criticizes in detail what he takes to be a misguided attempt and presents his own contribution to the topic in such a way as to make it the culmination and compendium of previous efforts. In the case of linear curves, instead, his subject needs consolidating, so Pappus' main focus is to present the curves as effective problem-solving tools, whose utility is proved by applying them to a number of constructions, and whose homogeneity is underscored by streamlining their definitions and the descriptions of their main properties.

In both cases, then, the past is appropriated for present purposes; the tradition is mined for results which are opportunely modified. In fact, as it becomes evident, especially in the case of the account on cube duplication, the tradition itself is not so much cut-and-pasted as tailor-made; it only comes to life when it takes on a certain guise to serve a particular purpose.

CHAPTER 5

THE INSIDE STORY

The time has come to try to get a clearer grasp of Pappus'
mathematical agenda. In the introduction we used the Arch of
Constantine as an analogy (of a sort) for Pappus' *Collection*.
The analogy can be further deployed to talk about style. In
the Arch the pieces from late antiquity look characteristically
different; although mathematics is remarkably conservative in
its modes of expression through the centuries,[1] and although
I am not in a position to define exactly what mathematical
style consists of (no more than one is capable of determining
exactly what artistic style consists of),[2] I think that there is
something in Pappus' *Collection* which is characteristically
different.

Some qualifications are in order: I do not wish to make a
case for Pappus' originality, nor restore him to his rightful
place in the heaven of great mathematicians, alongside Euclid
and Archimedes. It is more a matter of observing that, while
of course some things in mathematics do stay the same, different epochs produce different ways of doing mathematics. I
am aware that the evidence is not sufficient to enable me to
say that what I identify as Pappus' features were indeed his
own features, rather than the features of some intermediate
source which he used and we have now lost. Also, I am aware
that what I argue about some parts of the *Collection* is less in
evidence in other works by Pappus or in other parts of the
Collection itself (principally book 7). Pappus' work as a whole
is, however, incredibly diverse: I have already indicated that
different books of the *Collection* were geared to different audiences. Correspondingly, the emphasis put on certain features

---

[1] See e.g. Aujac (1984).
[2] See, on the difficulties of defining style, Ginsburg (1983).

rather than on others can be seen as strictly functional to the purpose of each text: the two commentaries and book 6 and 7 of the *Collection*, which are also in the nature of running complements to other texts, exhibit a much stricter relation to previous sources than the texts I have examined, where the appropriation of the past is more freely carried out.

5.1 The name of the game

The *Collection* contains several meta-mathematical or second-order pronouncements – passages where Pappus talks about mathematics instead of "doing" mathematics. Among them is a definition of analysis and synthesis, a definition of theorems and problems, and a division of problems and the procedures suitable for solving them into planar, solid and linear.

Quite a lot of work has been done on the definition of analysis and synthesis,[3] and I have already discussed the classification of problems and methods. Let us quote the remaining passage, which opens book 3:

Those who want to distinguish the things that are the object of research in geometry in a more expert way, my dear Pandrosion, think it appropriate to call problem that about which it is proposed to produce and construct something, while they consider theorem that in which, some things being assumed, what follows from them and above all what happens afterwards is investigated. Some of the ancients say that all things are problems, some that they are all theorems. Indeed, he who proposes a theorem, after surveying in whatever manner what follows, thinks it appropriate to investigate it this way and would not propose it rightly in another way. On the other hand, he who proposes a problem, [if he is ignorant and totally inexpert], even if he prescribes that something in a way impossible is to be constructed, should be excused and not blamed. In fact, it is the job of the person who is investigating to determine this: the possible and the not possible, and, if possible, when, in what way and in how many ways possible. But if there was someone pretending to be knowledgeable and proposing something in an inexpert way, he is not without blame.[4]

---

[3] E.g. Cornford (1932); Robinson (1936); Gulley (1958); Mahoney (1968); Gyekye (1972); Hintikka and Remes (1974), Szabó (1974), Rehder (1982); Behboud (1994).

[4] 30.1–17: Οἱ τὰ ἐν γεωμετρίᾳ ζητούμενα βουλόμενοι τεχνικώτερον διακρίνειν, ὦ κράτιστε Πανδροσίον, πρόβλημα μὲν ἀξιοῦσι καλεῖν ἐφ' οὗ προβάλλεταί τι ποιῆσαι καὶ

The distinction between problems and theorems does not originate with Pappus. We can compare his treatment to that of Proclus, who draws on several earlier sources: for instance, the followers of Zenodotus, who "used to distinguish theorem from problem in the sense that a theorem seeks to know what character is attributed to the matter it is investigating, whereas a problem asks under what conditions something exists."[5] Or Carpus and Geminus (both known to Pappus), who disagreed on the question of whether theorems are prior to problems or the other way round: "Carpus ... gives problems the priority in order, but Geminus judges primacy in terms of worth and perfection."[6] Or again the followers of Speusippus and Amphinomus (maybe a contemporary of Speusippus, early fourth century BC), who engage in the question of what is the "most appropriate designation for the objects of the theoretical sciences." They maintain that theorems are better because they do not "bring into being or ... make something not previously existing."[7] The mathematicians who followed Menaechmus, on the contrary, think that all mathematical enquiries are problems, but that sometimes the aim is "to provide something sought for, and at other times to see, with respect to a determinate object, what or of what sort it is, or what quality it has, or what relation it bears to something else."[8] As for Proclus, he thinks that "[b]oth parties are right. The school of Speusippus are right because the problems of geometry are of a different sort from those of mechanics ... [and] because theory is the predominant element in geometry, as making is

κατασκευάσαι, θεώρημα δὲ ἐν ᾧ τινῶν ὑποκειμένων τὸ ἑπόμενον αὐτοῖς καὶ πάντως ἐπισυμβαῖνον θεωρεῖται, τῶν παλαιῶν τῶν μὲν προβλήματα πάντα, τῶν δὲ θεωρήματα εἶναι φασκόντων. ὁ μὲν οὖν τὸ θεώρημα προτείνων συνιδὼν ὁντινοῦν τρόπον τὸ ἀκόλουθον τούτῳ ἀξιοῖ ζητεῖν καὶ οὐκ ἂν ἄλλοις ὑγιῶς προτείνοι, ὁ δὲ τὸ πρόβλημα προτείνων [ἂν μὲν ἀμαθής ᾖ καὶ παντάπασιν ἰδιώτης], κἂν ἀδύνατόν πως κατασκευασθῆναι προστάξῃ, σύγγνωστός ἐστιν καὶ ἀνυπεύθυνος. τοῦ γὰρ ζητοῦντος ἔργον καὶ τοῦτο διορίσαι, τό τε δυνατὸν καὶ τὸ ἀδύνατον, κἂν ᾖ δυνατόν, πότε καὶ πῶς καὶ ποσαχῶς δυνατόν. ἐὰν δὲ προσποιούμενος ᾖ τὰ μαθήματά πως ἀπείρως προβάλλων, οὐκ ἔστιν αἰτίας ἔξω.

[5] Proclus, *In Eucl.* 80.15–20.
[6] *op. cit.* 243.23–25.
[7] *op. cit.* 77.15–20.
[8] *op. cit.* 78.8. Cf. Bowen (1983); Mueller (1991).

in mechanics; every problem has also some theory in it."[9]
Note that the mathematicians of the school of Menaechmus
are not reported as saying that the problems of geometry
are of the same sort as those of mechanics – yet Proclus associates mechanics, problems and production without further
explanation.

On the issue of problems and mechanics, we also have the
opinion of Carpus the engineer, who

in his work on astronomy ... says that problems are prior in rank to theorems because problems discover the subjects whose attributes are under
investigation. And the enunciation of a problem ... only demands that
something clearly possible be done ... Now problems [here Proclus himself]
rightly do come before theorems in order of presentation, especially for
those who are coming to science from the arts concerned with sensible
things ... All of geometry, it appears, where it touches on the various arts,
operates by way of problems.[10]

Now, Pappus, too, is aware that some of the ancients
thought that all mathematical enquiries were theorems, while
some others reckoned that they were all problems, but he does
not discuss their views. It is clear that problems for Pappus are
about production, and not just statements of existence, but it
is also evident that he endorses neither radical opinion. In any
case, it is very significant that such issues should arise at all. It
seems that theorems and problems were actually perceived as
different, if complementary, kinds of knowledge, and that
making a stand for one or the other amounted to expressing a
strong view on the nature of mathematics.[11] In order better to
explain the distinction, let us read a passage from Aristotle's
On the soul entirely out of context. Aristotle is dealing with
the notion of what is the true definition of a thing and characteristically provides a geometrical example to illustrate his
meaning. Definitions, he says, should not just point at the

---

[9] Proclus, *In Eucl.* 78.13 ff. Proclus also reports the distinction between theorems
and problems by the followers of Posidonius (*op. cit.* 80.20–81.4).
[10] Proclus, *op. cit.* 241.18 ff.
[11] Cf. on this issue Zeuthen (1896); Rehder (1982); Knorr (1983) and (1986a), 353;
Napolitano Valditara (1988); Chihara (1990); Mueller (1991).

thing, but contain and exhibit its cause – for instance, there are two different answers to the question: "What is squaring a rectangle?" One is, the being equal of a square and a rectangle. The other one is, to find a mean proportional.[12] The first answer is correct and increases our information in that it sets an equivalence, yet it does not cover the entire spectrum of possible information on the issue. Namely, it does not produce either object, nor does it provide instructions as to how either object is to be produced. Aristotle's second answer can instead be translated into a set of instructions, whereby a square equal to the rectangle can be produced. In other words, one can do more than assert that such a mean proportional exists; one can specify how, given any rectangle, to find a square equivalent to it.

Again, of a theorem it is required that it be true – as long as people on the whole agree with the undemonstrable assumptions and with the demonstrative procedure, a theorem only needs to be clear and, at best, elegant (both concepts taken as more or less self-explanatory within localized mathematical practices). Problems, on the other hand, according to Pappus' definition, have to establish their own possibility first and then construct or *produce* the required item; therefore they are more dependent on the particular figure chosen In Pappus' words, with a problem one has to tell the when, the how and the in how many ways.

Pappus' strong concern for problems and construction;[13] the conditions he sets on the correct solution to a problem; his criticism of constructions which fail to determine the when, how and in how many ways; his preference for mechanical procedures; his interest in seeing how a construction or demonstration change when a certain element changes, all suggest

---

[12] Aristotle, *De anima* 413a13–20: οὐ γὰρ μόνον τὸ ὅτι δεῖ τὸν ὁριστικὸν λόγον δηλοῦν, ... ἀλλὰ καὶ τὴν αἰτίαν ἐνυπάρχειν καὶ ἐμφαίνεσθαι.... οἷον τί ἐστιν ὁ τετραγωνισμός; τὸ ἴσον ἑτερομήκει ὀρθογώνιον εἶναι ἰσόπλευρον.... ὁ δὲ λέγων ὅτι ἐστὶν ὁ τετραγωνισμὸς μέσης εὕρεσις. Knorr (1986a), 24 cites the passage as relevant to the issue of reduction of one problem to another.

[13] Cf. Bos (1996).

a type of mathematical practice which could be characterized by the general term of "manipulation."[14] In the next sections I will describe what I take to be the most distinctive ways in which Pappus' manipulations are carried out.

Generalizations

By generalization I mean a case where Pappus takes a statement and expands it to include a wider class of geometrical objects. One example is Pappus' modification, in book 4, of the theorem of Pythagoras, known to the ancients as part of Euclid's *Elements*.[15]

Just to remind the reader: according to the Euclidean formulation, in any right-angled triangle the square on the hypotenuse is equal to the sum of the squares on the other two sides. Pappus' theorem refers instead to *any* triangle, not necessarily a right-angled one. The proof of the theorem is rather straightforward, in that it employs simple properties of parallels. Pappus himself adds at the end: "this is much more general than the one proved in the *Elements* about squares in right-angled triangles."[16] Had the previous theorem *not* been famous, Pappus could not have presented his own contribution as one step further than a well-known piece of geometry. The original on which the generalization is carried out must already be part of some tradition for the entire operation to produce its best benefits.

This is the case again with a theorem which we find both in book 5 and in Archimedes' *Sphere and Cylinder I*. Archimedes has three separate propositions to establish, respectively, the area of a sphere and of a spherical segment greater than, or

---

[14] The references to Pappus which we find in Proclus chiefly refer to this type of operation: *In Eucl.* 189.12; 197.6; 249.20–250.12. These are, respectively, the demonstration that the converse of a certain postulate is not true; the addition of some axioms and the reduction of a proof so that it does not require a supplementary construction.

[15] Pappus at 176.9–178.11; the "theorem of Pythagoras" in Euclid, *El.* I.47. Cf. Proclus *ad loc. In Eucl.* 429.15. The proposition at *El.* VI.31 is a different generalization of the theorem of Pythagoras.

[16] 178.11–13: καὶ ἔστι τοῦτο καθολικώτερον πολλῷ τοῦ ἐν τοῖς ὀρθογωνίοις ἐπὶ τῶν τετραγώνων ἐν τοῖς στοιχείοις δεδειγμένου.

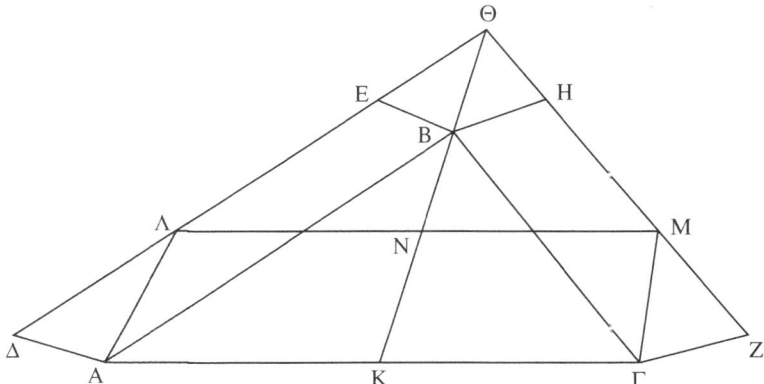

Fig. 5.1. If we construct parallelograms on two sides of the triangle ABΓ, i.e. ΛABΘ and BΘMΓ, and prolong their sides, the sum of the parallelograms ABEΔ and BΓZH is equal to the parallelogram contained by AΓ and ΘB under the angle ΛAΓ which is the sum of the angles BAΓ and ΔΘB.

less than, a hemisphere. Pappus has one single theorem that concerns *any* spherical segment, where the sphere is seen as a limit-case.[17] The demonstrative procedures are along similar lines, all proofs being by *reductio ad absurdum*; Pappus' generalization is made possible by an accurate choice of auxiliary lemmas. The first of those establishes the area of a solid obtained by the rotation around the diameter of a semi-circumference of a polygon inscribed in a portion of that semicircumference; next comes a lemma which expands on the former by considering the limit-case of the area of a solid produced by the rotation of a polygon inscribed this time in a whole semicircumference. Analogous results are then proved for polygons circumscribed to semicircumferences, and for polygons circumscribed to a portion of semicircumference.[18] Thus, having explored all the possible combinations, the final proposition, which leaps to curved space, comes as a natural next step. Once again, there is a famous reference point, which

[17] 382.19–386.21 and Archimedes, *Sph. Cyl. I* 35, 48 and 49, respectively. See Heath (1921), II 213.

[18] 366.11 ff.; 368.25 ff.; 370.1 ff. and 372.16 ff., respectively.

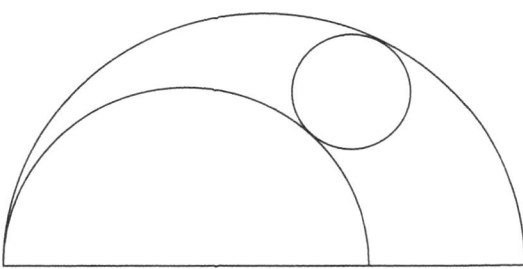

Fig. 5.2. First configuration.

Pappus acknowledges, but from which he also distances himself: "[those things], as we have said, have been proved by Archimedes and we will prove them differently."[19]

Particular cases

Particular cases are in a sense the inverse of generalizations: a proposition is sub-divided according to how a change in some of the elements, or a change in the relation between the geometrical objects in question, affects the demonstration or construction. An example will make this clearer: book 4 of the *Collection* presents a whole series of theorems which are sub-divided, both for the enunciation and for the proof, into particular cases.[20] At issue is the relation between two semicircles and one circle tangent to one another – this configuration comprises three possibilities, which depend on how one constructs the diagram. The circle and one semicircle can be contained within the second semicircle (first configuration), or both semicircles can be contained within the circle (second configuration), or all figures can be external to each other (third configuration).

It is remarkable that all these various possibilities arise

[19] 362.18–20: τὰ δ' ὑπὸ τοῦ Ἀρχιμήδους, ὡς εἴρηται, δειχθέντα καὶ ἄλλως ἀποδείξομεν. As we have seen in chapter two, this result could have been taken from Hero's *Metrica*, or from a source common to Pappus and Hero, but what is of significance is that Pappus chose to modify Archimedes' result, i.e. he modified traditional results in a certain way.

[20] 212.22–218.11.

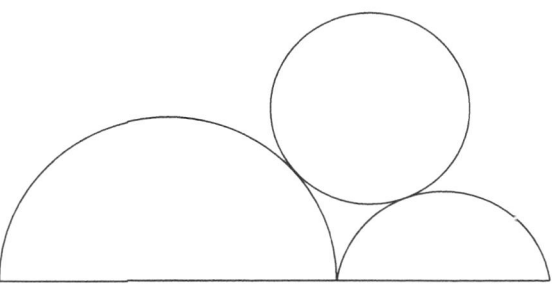

Fig. 5.3. Third configuration.

from different arrangements of the diagram, not from different geometrical relations between the figures involved. Pappus refers to the "first", "second" and "third" diagram and quite often in the account points out the relative variations in their proofs.[21] As next step he considers two semicircles and *two* circles, which can be arranged in three different ways, and then, in a sort of mathematical juggling, three circles and three semicircles. In this latter configuration, four sub-cases have to be considered.[22]

An analogous approach is found in book 5 of the *Collection*, where, having constructed two isosceles triangles whose bases are unequal but whose other sides are all equal to each other, it is required to construct on the same bases two other similar and isosceles triangles, so that the sum of their equal sides is equal to the sum of the equal sides of the first two triangles. Pappus does this by positing two more triangles equal to the two first triangles one is required to construct – the advantage being that these two new triangles (ΠΡΣ, which is equal to ΓΖΔ, and ΠΡΤ, which is equal to ΑΕΒ) can be juxtaposed in the second, third and fourth diagram to the triangles which it is required to construct, making the result visually more evident.

Zenodorus, from whom this puzzle is drawn, only considers the case where the first triangle is greater than the second one.

---

[21] Diagram is here καταγραφή. E.g. 214.1; 216.1–2, 10.
[22] 218.12–224.11 and 224.12–230.8, respectively.

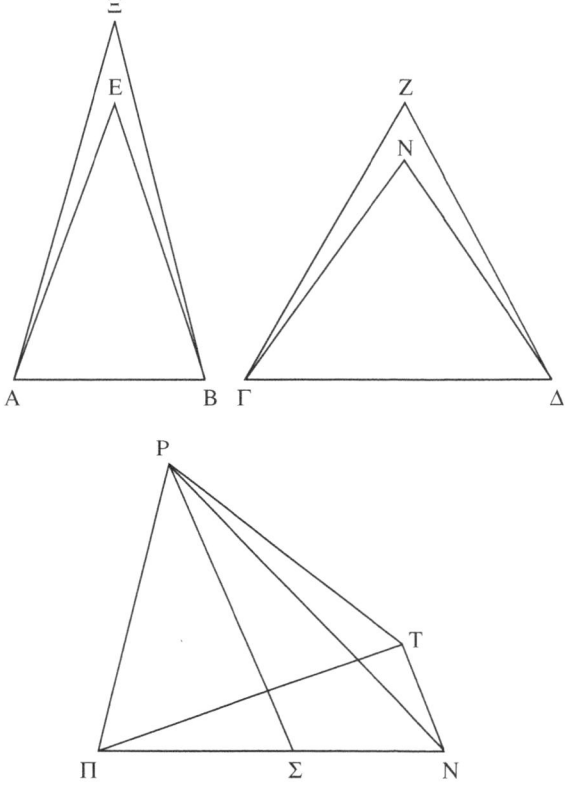

Fig. 5.4. AEB is greater than ΓΖΔ.

Pappus, on the other hand, examines three subcases: the first triangle is greater than the second one; the first triangle is smaller than the second one; the two triangles are equal (this latter is a limit-case, because it had been assumed that their bases were unequal).[23]

Among the qualities required of a mathematician, Proclus (quoting Aristotle) indicates "certain standards of judgment. In the first place, he should know when he can make his demonstrations general and when he must look to the prop-

[23] 330.6–332.10.

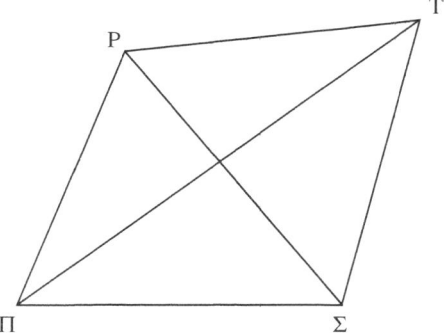

Fig. 5.5. AEB is equal to ΓΖΔ.

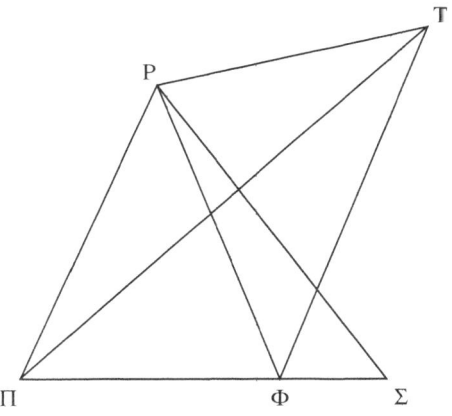

Fig. 5.6. AEB is smaller than ΓΖΔ.

erties of the species." And again: "A case announces that there are different ways of making the construction, by changing the position of the points, lines, planes, or solids involved. Variations in case are generally made evident by changes in the diagram, wherefore it is called case, because it is a transposition in the construction."[24] I take Proclus as

---

[24] Proclus, *In Eucl.* 32.23–33.2; 212.5–11, respectively. At 269.21–25 he criticizes "the school of Philon" for a version of Euclid's *Elements* 1.8 which deals with subcases: "although Philon's procedure is elegant, its use of a variety of cases makes it unsuited for an elementary treatise." See also Cambiano (1985).

evidence that the issues of manipulation of results along the lines of generalization and examination of sub-cases, although they were by no means new,[25] existed as issues in late antiquity and were given a particular spin by the fact that often they were manipulations conducted on material handed down from the tradition; they were a way of appropriating the past.

Arithmetizations

What I call arithmetization consists in assigning specific numbers to a general mathematical proposition, which has already been established, and then in formulating it again. This operation appears to be a simple rehearsal of what has already been attained – a sort of exercise for the reader, who can thus further convince himself that the proof or the construction in question works. Something similar is found in Hero of Alexandria's *Metrica*, which deals with areas and volumes of geometrical figures. In the great majority of the cases, geometrical objects are introduced in the *Metrica* with specific numbers for their dimensions (for instance, "Let ABΓ be a scalene triangle, that has AB of 13 units, BΓ of 11 units and AΓ of 20 units"),[26] and propositions mix or juxtapose general demonstrations (for instance, about the area of any triangle) and specific calculations of the area, say, of that particular triangle. In other words, a formulation of the area or volume of a geometrical figure will be presented together with the calculation of the area or volume of a *specific* geometrical figure – of one item of the species, as it were. The geometrical part is seen by Hero as a demonstration, whereas the procedure with the numbers is sometimes characterized as an "apprehension" or a "grasping", often as a "synthesis", but not simply as an example.[27]

---

[25] Euclid's *Elements* contains of course many propositions which are divided into sub-cases, e.g. 3.25; 3.33; 3.35; 3.36; 4.5; 5.8.

[26] Hero, *Metrica*, 14.18–20: "Εστω τρίγωνον ἀμβλυγώνιον τὸ ΑΒΓ ἔχον τὴν μὲν ΑΒ μονάδων ιγ, τὴν δὲ ΒΓ μονάδων ια, τὴν δὲ ΑΓ μονάδων κ.

[27] Hero, *op. cit.* 118.24–26 and 38.26–27; 48.24; 54.2; 56.13–14; 58.9. It is significant that we never find expressions like λόγου χάριν ('for the sake of argument, "for example"') in connection with the numbers. See also Vitrac (1994).

In book 2 of the *Collection*, Pappus provides a series of rules for multiplication of units by multiples of ten and of one hundred. Many of the enunciations of the propositions actually maintain that their aim is to "tell" the product of those numbers without multiplying them.[28] Most propositions are supplemented by applications to specific numbers, that is, some numbers are taken which satisfy the requirements and then it is shown how the procedure works in the particular case of those numbers. The final proposition is a sort of number game involving two short epigrams, where letters are substituted by the corresponding numbers (the Greek signs for numbers were the same as for letters)[29] and multiplied by each other.

I will quote one proposition in full as an example:

Let a multitude ⟨of numbers⟩ be the multitude labelled by the As, of which each is less than a hundred and divisible by ten, and another multitude of numbers, the multitude labelled by the Bs, of which each is less than a thousand and divisible by a hundred, and let it be required to tell the product of the As and Bs without multiplying them. Let then basic numbers of the As be the ⟨numbers⟩ labelled by the Hs, the units 1, 2, 3 and 4, and let basic numbers of the Bs be the ⟨numbers⟩ labelled by the Θs, the units 2, 3, 4 and 5, and assuming that the product of the basic numbers, [of 2, 3, 4, 2, 3, 4, 5] that is, E, is of 2880 units, let the multitude of numbers labelled by the As added to twice the multitude of numbers labelled by the Bs be first divided by four [with the result Z, it in fact divides them]. And Apollonius proves that the product of all the As and Bs is of as many myriads as there are units in E, with the same name as the number Z, that is 2880 triple myriads. [In fact one myriad with the same name as the Z, that is triple, multiplied by the E, that is 2880, makes the number of the products of the As and the Bs. But indeed the product of the numbers labelled by the As and Bs is as many myriads as there are units in E, with the same name as the number Z.]. But let the multitude of numbers labelled by the As, added to double the multitude of numbers labelled by the Bs, divided by four first have the remainder of one. And Apollonius gathers that the product of the numbers labelled by the As and Bs is as many myriads with the same name as the Z, as is ten times the E; but if the aforesaid quantity divided by four

---

[28] The usual formula is: "and let it be required to tell the product of those [numbers] without multiplying them" (καὶ δέον ἔστω τὸν ἐξ αὐτῶν στερεὸν εἰπεῖν μὴ πολλα- πλασιάσαντα αὐτούς).

[29] A good description of the Greek mathematical notation system in Heath (1921), I ch. 2.

has the remainder of two, the product of the As and Bs is as many myriads with the same name as the Z, as is a hundred times the number E; if the remainder is three, the product of those numbers is equal to as many myriads with the same name as the Z as is a thousand times the number E.[30]

According to Thomas Heath's reconstruction, book 2 assumes a basic rule that splits "the multiplication of any number of factors, each of which is one or other of the following: (a) a number of units as 1, 2, 3 … 9, (b) a number of tens as 10, 20, 30 … 90, (c) a number of hundreds as 100, 200, 300 … 900" into "the separate multiplication of (1) the bases, πυθμένες, of the several factors [the bases are obtained by dividing each multiple of a hundred by a hundred, each multiple of ten by ten and by taking the units simply as they are] and (2) the powers of ten contained in the factors."[31] Given that there is the condition that each multiple of a hundred must be less than a thousand, that each multiple of ten must be less than a hundred and that each unit must be less than ten, any πυθμέν will be less than ten. The first multiplication, i.e. the multiplication of the bases, would thus seem to pose no particular problem, and it in fact is treated as unproblematic in the text.

---

[30] 6.6–8.11: "Εστω πλῆθος ἀριθμῶν τὸ ἐφ' ὧν τὰ Α, ὧν ἕκαστος ἐλάσσων μὲν ἑκατοντάδος μετρούμενος δὲ ὑπὸ δεκάδος, καὶ ἄλλο πλῆθος ἀριθμῶν τὸ ἐφ' ὧν τὰ Β, ὧν ἕκαστος ἐλάσσων μὲν χιλιάδος μετρούμενος δὲ ὑπὸ ἑκατοντάδος, καὶ δέον ἔστω τὸν ἐκ τῶν ἐφ' ὧν τὰ Α Β στερεὸν εἰπεῖν μὴ πολλαπλασιάσαντα αὐτούς. Ἔστωσαν γὰρ πυθμένες τῶν μὲν ἐφ' ὧν τὰ Α οἱ ἐφ' ὧν τὰ Η, μονάδες α' καὶ β' καὶ γ' και δ', τῶν δὲ ἐφ' ὧν τὰ Β οἱ ἐφ' ὧν τὰ Θ, μονάδες β' και γ' και δ' καὶ ε', καὶ ληφθέντος τοῦ ἐκ τῶν πυθμένων στερεοῦ [τῶν β' γ' δ' β' γ' δ' ε'], τουτέστιν τοῦ Ε, μονάδων ὄντος, βωπ', τὸ πλῆθος τῶν ἐφ' ὧν τὰ Α προσλαβὸν τὸν διπλασίονα τοῦ πλήθους τῶν ἐφ' ὧν τὰ Β μετρείσθω πρότερον ὑπὸ τετράδος [κατὰ τὸν Ζ, μετρεῖ δὲ αὐτούς]. καὶ δείκνυσιν ὁ Ἀπολλώνιος τὸν ἐκ πάντων τῶν ἐφ' ὧν τὰ Α Β στερεὸν μυριάδων τοσούτων, ὅσαι εἰσὶν ἐν τῷ Ε μονάδες, ὁμωνύμων τῷ Ζ ἀριθμῷ, τουτέστιν τριπλῶν μυριάδων, βωπ'. [μία γὰρ μυριὰς ὁμώνυμος τῷ Ζ, τουτέστιν τριπλῆ, ἐπὶ τὸν Ε, τουτέστιν τὰ, βωπ', γενομένη ποιεῖ τὸν ἐκ τῶν στερεῶν ἀριθμὸν τὸν ἐφ' ὧν τὰ Α Β· ὁ ἄρα ἐκ τῶν ἀριθμῶν στερεὸς τὸν ἐφ' ὧν τὰ Α Β μυριάδες εἰσὶν τοσαῦται, ὅσαι εἰσὶν ἐν τῷ Ε μονάδες, ὁμώνυμοι τῷ Ζ ἀριθμῷ.] Ἀλλὰ δὴ τὸ πλῆθος τῶν ἐφ' ὧν τὰ Α, προσλαβὸν τὸν διπλασίονα τοῦ πλήθους τῶν ἐφ' ὧν τὰ Β, μετρούμενον ὑπὸ τετράδος καταλειπέτω πρότερον ἕνα. καὶ συνάγει ὁ Ἀπολλώνιος ὅτι ὁ ἐκ τῶν ἀριθμῶν ἐφ' ὧν τὰ Α Β στερεὸς μυριάδες εἰσὶν τοσαῦται ὁμώνυμοι τῷ Ζ, ὅσος ἐστὶν ὁ δεκαπλασίων τοῦ Ε, ἐὰν δὲ τὸ προειρημένον πλῆθος μετρούμενον ὑπὸ τετράδος καταλείπῃ δύο, ὁ ἐκ τῶν ἀριθμῶν στερεὸς τῶν ἐφ' ὧν τὰ Α Β μυριάδες εἰσὶν τοσαῦται ὁμώνυμοι τῷ Ζ, ὅσος ἐστὶν ὁ ἑκατονταπλάσιος τοῦ Ε ἀριθμοῦ, ὅταν δὲ τρεῖς καταλειφθῶσιν, ἴσος ἐστὶν ὁ ἐξ αὐτῶν στερεὸς μυριάσιν τοσαύταις ὁμωνύμοις τῷ Ζ, ὅσος ἐστὶν ὁ χιλιαπλάσιος τοῦ Ε ἀριθμοῦ.

[31] Heath (1921), I 55. See also Nesselmann (1842); Tannery (1880a).

As for the second multiplication, i.e. the multiplication of the powers of ten contained in the factors, it is generally carried out thus: one counts how many numbers there are in each group and then doubles this quantity in the case of multiples of a hundred, sums the numbers obtained from the various groups and divides the total by four. If the total divides by four exactly, the result of the second operation will be one myriad (ten thousand) to the power of the quotient of the division. If the remainder is one, two or three, the product will be one myriad to the power of the quotient of the division, further multiplied by ten, one hundred or one thousand, respectively.

Our extant version of book 2 contains no demonstration of this basic multiplication rule. All we find are references to a linear proof (τὸ γραμμικόν or διὰ τῶν γραμμῶν) associated with Apollonius, and applications of the rule to specific numbers.[32] It is not clear what this "graphic" demonstration may have looked like, but it is referred to with verbs such as δείκνυμι and ἀποδείκνυμι, which in Pappus and elsewhere usually denote a rigorous proof. What is more relevant for us here is that eleven out of the twelve variations on the multiplication rule that form book 2 are rehearsed "by means of numbers." Although not presented as a proof (it is usually introduced by the weak demonstrative clause "it is clear by means of numbers" – φανερόν ἐστι διὰ τῶν ἀριθμῶν), the numerical procedure is more than a series of examples. For instance, props. 16, 18, 19, 20 and 22 show that the rule is not applicable in all cases, precisely by evidencing that in some *particular* cases it does not apply. On the whole, the repetition itself of the application of the general rule to sets of numbers suggests that the exercise was as much an integral part of the persuasive process as the proof; by allowing the reader to

---

[32] 4.3–4; 6.4–5; 8.27–28; *passim.* Pappus' commentary on Ptolemy contains rules for sexagesimal division; see Mogenet (1951). Cauderlier (1978), 52, mentions a multiplication table (maybe from the sixth century AD) from Antinoopolis which contained similar exercises, and Eutocius, *In Archimedis Dimensionem Circuli*, 300.15 ff. mentions multiplications and divisions of myriads which would be familiar to someone who had gone through Magnus' *Logistics*.

manipulate the numbers at will, it facilitated his access to this piece of mathematical knowledge.

"Arithmetizations" can also help establish the conditions for the validity of a construction. We have seen how Pappus criticizes the anonymous solution to the problem of the two mean proportionals because (among other things) its author fails to determine the position of a certain point, which varies according to a certain ratio. Pappus' way out of this particular difficulty consists in attributing specific values to the ratio, and in subsequently working out in each case within what interval the point will be situated. He starts with the simplest ratio (2:1) and proceeds to 5:1 and above, stating the corresponding positions of the point in the diagram.

Numbers appear again in book 3, in the context of the treatment of various kinds of means, which follows the account of cube duplication. Pappus specifies the set of least numbers which satisfy the conditions for each particular type of mean, and provides a complete table (πλινθίον) of these numbers at the end of the account, "for the sake of accessibility."[33] There is also a fully-worked out application of numbers to one of the propositions, which is introduced as an example.[34]

Pappus' arithmetizations can be applied to "strange" figures. We have already pointed out the final proposition of his treatment of the spiral, where he measures the inner area of some portions of the surface contained by the curve and by its generating line, and comes to the conclusion that they have the ratio 4:3:2:1 to each other. This can be seen as a step towards greater knowledge of the spiral: by giving a number to its area, Pappus extends its mathematical description.[35]

In book 8, Pappus concludes his account of the problem of the inclined plane with the words:

The geometrical analysis of the problem has been proved above, now, in order to produce the construction and the proof as an example, let the

---

[33] 100.19–104.13: τοῦ προχείρου χάριν. The more usual word for a table of this sort, e.g. for Ptolemy's *Handy Tables*, was κανών (see Pappus himself at 48.16).

[34] 78.18–80.23.

[35] 242.1–12.

weight A be 200 talents, for instance, and let it be dragged along the plane of the horizon by the moving power Γ, that is, let the moving be 40 men; let the angle KMN, [i.e. the angle of inclination of the plane] be two-thirds of a right angle.[36]

In conclusion, if it takes 40 men to move a weight of 200 talents along a horizontal plane, it will take 300 men to move the same weight up a plane inclined to the horizon by an angle equal to two-thirds of a right angle. The first thing to remark here is that, if we stick to Pappus' own definition of analysis and synthesis as given in book 7 of the *Collection*, what precedes the "example" is a *synthesis*, not an analysis. On the other hand, if calling the general proposition an analysis amounts to saying that the numerical example is a synthesis, then Pappus may be sharing Hero's terminology in the *Metrica*. What the παράδειγμα in fact amounts to is a repetition of the argumentation, but this time applied to specific numbers for the weight, the angle and the force, which constitute the data of the problem. Although not counting as a rigorous demonstration, the exercise satisfies the requirements usually set for a construction and, while evidently relying on the general proof provided immediately before it, it has a deductive structure of its own and its own appeals for validity – it quotes the canon on straight lines in a circle "as contained in Ptolemy," which it does not do in the course of the main proof.[37]

Again in book 8 we find a string of theorems on the properties of cog-wheels.[38] The main result is formulated first as a theorem, that the ratio between the number of the teeth on two cog-wheels is the same as the ratio between their diameters, and then as a problem: given a wheel with a given number of teeth, and having applied another wheel with a given number of teeth to the first one, find the diameter of the sec-

---

[36] 1056.30–1058.5: Ἡ μὲν οὖν γεωμετρικὴ τοῦ προβλήματος ἀνάλυσις ὑποδέδεικται, ἵνα δὲ καὶ ἐπὶ παραδείγματος ποιησώμεθα τήν τε κατασκευὴν καὶ τὴν ἀπόδειξιν, ἔστω τὸ μὲν Α βάρος ταλάντων, εἰ τύχοι, σ’ ἀγόμενον ἐν τῷ παραλλήλῳ ὁρίζοντι ἐπιπέδῳ ὑπὸ τῆς Γ κινούσης δυνάμεως, τουτέστιν οἱ κινοῦντες ἔστωσαν ἄνθρωποι μ, ἡ δὲ ὑπὸ ΚΜΝ γωνία, τουτέστιν ἡ ὑπὸ ΕΘΛ, διμοίρου ὀρθῆς.

[37] 1058.12–14.

[38] 1102.11 ff.; the formulation as a problem at 1106.26 ff.

ond wheel. Both with the theorem and the problem, the general statements are applied to numbers, the same numbers in both cases. In both theorem and problem, the numbers clearly amount to an example: they are there as if to confirm the actual feasibility of the cog-wheel device, to provide a sense of reality. Now, Hultsch expunges some of the passages where the numbers are given,[39] on the grounds that "the demonstration of this problem is general; therefore the determinate numbers are foreign to the author's purpose."[40] Yet the terminology itself suggests that those numbers can be seen as the link between a material quantity (the "multitude" of the teeth themselves, πλῆθος) and the same quantity expressed as a number (ἀριθμός).[41] Thus, far from containing interpolations, this passage is meant to stress the interaction of mathematics and mechanics, in perfect accord with Pappus' intentions in book 8.

5.2 The sins of the fathers

Pappus has been described as completely devoted to the cult of the past. I have argued that his attitude is more complicated than that: he appropriates the past for his own purposes, and modifies previous results not at random, or as availability of sources dictates, but in deliberate ways that can be observed and described. Of course, when we consider manipulations of the kind I have outlined, the presence of a mathematical tradition looms, if anything, even larger. It is fundamental to the possibility of manipulations at all, and of major importance in their acquiring significance. The complexity of Pappus' attitude to the cult of the past will be explored in the next sections.

---

[39] 1108.4 ff.
[40] 1108n: "demonstratio huius problematis generalis est; ergo alieni a scriptoris ratione sunt numeri definiti;" the passages concerned are 1108.4, 6, 14–15, 16, 18–21.
[41] Cf. 1104.1–2 and 1108.3 ff. A discussion about different terms for number in Klein (1934–6).

The passage from book 4 we quoted in the previous chapter accused both Archimedes and Apollonius of a major fallacy:

it seems to the geometers that it is no small mistake when [the construction of] a planar problem is discovered by someone by means of conics or of linear [curves], and in general when it is solved by means of a kind not its own, as is the case in the fifth book of Apollonius' conics with the problem about the parabola and in Archimedes' spirals about the solid neusis applied to the circle.

At the end of book 4, and following a practice we have seen in book 3 as well, Pappus sets out to "correct" the mistake and provides an alternative to Archimedes' fallacious construction. The proposition is preceded by, and depends on, two lemmas involving a hyperbola and a parabola, and is followed by still another version of the objections against Archimedes.[42] This last part, which is also the conclusion of book 4, has unfortunately not come down to us in its entirety. It repeats: "some geometers accuse Archimedes of having used a solid problem not as he ought to † (they) prove ⟨how⟩ to find ⟨a straight line equal⟩ [these words have been reconstructed by Hultsch] to a circumference by means of planes, using the theorems mentioned about the spiral."[43]

Lacunae are not the only problem of interpretation. In his works as they have come down to us, Archimedes did not explicitly link his study of the spirals to the quadrature of the circle.[44] According to Iamblichus *apud* Simplicius, Archimedes was indeed among the people who "put together" a square equal to a given circle, but the text, as we have seen, is not clear as to what procedure was employed. Since the other geometers mentioned by Iamblichus in the relevant passage all employ linear curves of various kinds, some contemporary scholars have interpreted the passage as referring to the spiral. The reliability of this testimony is of course somewhat marred

---

[42] The whole passage at 298.11–302.12.

[43] 302.14–18: αἰτιῶνται δὲ αὐτοῦ τινες ὡς οὐ δεόντως χρησαμένου στερεῷ προβλή-ματι † δεικνύουσιν ὡς καὶ διὰ τῶν ἐπιπέδων εὑρεῖν ἔστιν εὐθεῖαν ΄σην τῇ τοῦ κύκλου περιφερείᾳ χρησάμενον τοῖς ἐπὶ τῆς ἕλικος εἰρημένοις θεωρήμασιν.

[44] *Pace* Knorr (1986a), 167–70.

by the fact that the discovery of the area of the circle is attributed to a Sextus the Pythagorean.

The question then remains, who are the "geometers" who are reported as accusing Archimedes, and on what did they base their accusations?[45]

In fact, the cumulative effect of the two passages in book 4 is that Archimedes is accused of two separate things, both located with prop. 18. One has to do with squaring the circle. As is well known, Archimedes had established the classical formulation for the area of the circle. He must have convinced his contemporaries of its correctness, since the result is often quoted, and its authorship was retained, by the later tradition.[46] Yet we also find appreciations of the fact that once the theorem was established, i.e. the equivalence between a circle and a certain triangle, the question remained of how this triangle was to be produced. Uneasiness with the state of the question regarding the quadrature of the circle is expressed by several late ancient commentators. Marinus defines as "feasible" what it is possible to produce and construct (for instance, given two points, the drawing of a circle through them); squaring the circle is adduced as an example of "non-feasible" (ἄπορον), or rather feasible but not actually constructed, (ποριστόν rather than πόριμον); it is also termed "ordered", (τεταγμένον), but "not known" (ἄγνωστον).[47] Again, Eutocius comments that it may occur to somebody that positing a line equal to a circumference is problematic, because the fact that such a line exists has neither been proved by Archimedes (on whose *Measurement of the circle* Eutocius is commenting) nor transmitted by anyone else. But that such a line exists is evident, he continues, and there should be no need to argue that Archimedes made no mistakes; indeed, he is admirable for providing such a clear and easy solution to such difficult problems. But the critics of Archimedes rear their heads again

---

[45] Knorr (1986a), 326 ff. has suggested that Pappus' source for the last part of book 4 and for the criticism of mixed procedures on the whole may be Aristaeus the Elder's *Solid Loci*, mentioned in book 7.

[46] E.g. in Pappus 258.17–19 and 312.25 ff. Other examples cited by Knorr (1986a), 83n109.

[47] Marinus, *In Euclidis Data* 240.18–21 and 242.13–14, respectively.

later in the text, when Eutocius is discussing the approximations of the ratio between diameter and circumference. He has to admit that other people have been more accurate than the Syracusan geometer, but too much accuracy, claims Eutocius, is not appropriate to the aim of the book, which is to find an approximation suitable for the uses of life. One of the people who criticized Archimedes for not having expressed with accuracy the nature of a line equal to the circumference of the circle was Sporus.[48] We have seen Sporus' objections to the quadratrix and it is conceivable that similar problems had been raised regarding the spiral as applied to the squaring of the circle. The common denominator, for which Eutocius provides further evidence, seems to be Sporus' uneasiness with determination in constructions which deal with the relation between straight lines and circumferences. Pappus may have been put in a difficult position by such objections. He salvages both quadratrix and spiral by showing their applicability to problems such as the division of an angle by any ratio; in the case of the spiral he also avoids prop. 18 and reaffirms the value of the *Spirals* as an example of wondrous doctrine.[49]

The other charge against Archimedes is that he mixes categories. In fact, Pappus' alternative neusis construction itself employs solid procedures for a planar problem, so that the criticism could easily be retorted against him. And again the cochloid, a *linear* curve, is applied by Pappus in a previous treatise (so he says) to the trisection of an angle (a *solid* problem) and, here and in book 3, to the duplication of the cube (another *solid* problem). In book 4, moreover, Pappus says that the problem of the angle-trisection has been *proved*

---

[48] Eutocius, *In Dim. Circ.* 266.15–268.8 and 300.15 ff., respectively. Sporus apparently says in the Κηρία that his teacher, Philo of Gadara, had a better approximation than Archimedes. After repeating that these people have missed the point of the *Measurement of the circle*, Eutocius comments that for their approximations they use multiplications and divisions of myriads which are difficult to follow, unless one has been through the *Logistics* of Magnus. See also Ammonius, *In Aristotelis Categorias* 75; Simplicius, *In Aristotelis Physicam* 59 f.; 1082 (quoting Ammonius) – they both wonder whether it is at all possible to find a line equal to a circumference. Knorr (1986a), 361 ff. contains most of these references; see also Goldstein (1989).

[49] 234.3: θαυμαστὴ ἐπιβολή.

to be solid; the generalized angle-division is a linear problem, and, in its turn, it has been *proved* so.[50] Pappus' remarks sound as if there was a rigorous demonstration that certain problems belonged to one category rather than another, but, on modern criteria, such a proof should also establish the impossibility of the contrary, i.e. it should prove that a certain problem can be solved by solid procedures *and* that it cannot be solved by planar procedures. A proof satisfying these criteria was first produced in the nineteenth century, and it is unlikely that Pappus knew an equivalent formulation, or indeed that he saw those criteria as necessary or binding. From his words, it sounds as if the fact that the trisection of an angle is a solid problem is proved by producing a solution that employs conic sections, and the fact that the division of arcs and angles according to any ratio is a linear problem is shown by two proofs which employ quadratrix and spiral.[51] In all cases we have *individual* instances of valid solutions; these must have counted as a *general* demonstration.[52]

Moreover, Pappus is putting into the linear category objects which can be ascribed to that category only on *a posteriori* considerations. That is, a curve belongs to the category of linear only after it has proved effective when applied to linear problems, to which it can only be (legitimately) applied if it belongs to that category in the first place. One could at the same time accept a solution as valid and use it as a platform for methodological directions which, if strictly followed, would have prevented that solution from being found at all. On the face of it, we have a textbook case of circular reasoning and a clear case of double standards, as scholars have not failed to notice:[53] category-crosses are tolerated when Pappus himself is responsible for them.

---

[50] 284.21–24: Τὸ μὲν οὖν τὴν δοθεῖσαν γωνίαν ἢ περιφέρειαν τρίχα τεμεῖν στερεόν ἐστιν, ὡς προδέδεικται, τὸ δὲ τὴν δοθεῖσαν γωνίαν ἢ περιφέρειαν εἰς τὸν δοθέντα λόγον τεμεῖν γραμμικόν ἐστιν καὶ δέδεικται μὲν ὑπὸ τῶν νεωτέρων.

[51] 272.7–14 and 284.21–24.

[52] Cf. Seidenberg (1966); Étienne & Roels (1986). A proof that the problem of the two mean proportionals could not be solved by planar methods was first given by Wantzel (1837).

[53] Knorr (1989), 69–70.

I think that the ambiguities of Pappus' criticisms of Archimedes in book 4 can be put down to a compromise (in my opinion, not very well negotiated) between conflicting demands: a desire to stick to the classification and a desire to produce solutions to problems which satisfy the criteria laid out at the beginning of book 3. If anything, the second demand seems more pressing than the first one. Prop. 18 of Archimedes' *Spirals* is rejected on both accounts, but, when Pappus tries to make good Archimedes' alleged mistake, he himself breaches the category rule, while arguably producing an alternative which would be in line with the second demand. Consider the category-cross between linear and solid: linear curves provided a means for a general solution (not just the trisection) to the division of angles and arcs: given the conditions of use, one might well privilege the most comprehensive method. Also, in the case of the cochloid, the existence of a standard instrument to draw it, described by Eutocius and known to Pappus, probably determined its success, especially given that conic sections were notoriously cumbersome to draw. Even centuries later, Eutocius remarks: "given the difficulty of instruments, it does not seem useless to the writers of mechanics often to draw the conic sections in a plane by continuous points," and he mentions, as something worth mentioning, a device invented by his master Isidorus for drawing the parabola. Thus, when Pappus emphasizes at least twice the difficulty of drawing conic sections, this is not an insignificant detail.[54] While the use of conic sections for neusis constructions was so well established that they are practically the only procedure Pappus employs in book 4 (even when the neusis is planar),[55] where other types of constructions were concerned the use of conic sections could effectively hinder the constructability and determinability of a problem. Pappus' indulgence towards some category-crosses is then to

---

[54] Pappus at 54.26 f.; 1070.9 f. Eutocius, *In Sph. Cyl.* 84.8–11 and *In Apol. Con.* 230.27–232.5: οὐκ ἄχρηστον φαίνεται τοῖς τὰ μηχανικὰ γράφουσι διὰ τὴν ἀπορίαν τῶν ὀργάνων καὶ πολλάκις διὰ συνεχῶν σημείων γράφειν τὰς τοῦ κώνου τομὰς ἐν ἐπιπέδῳ.

[55] Apart from the final passage, see 272.7 ff.

be explained in terms of the concern about what constitutes a valid solution to a problem. The requirement that a problem should be constructed and its construction determined satisfactorily overrides the prescription emanating from the classification.

This interpretation, if plausible, can help us understand another difficulty: where do mechanical procedures fit in? Unlike procedures via conic sections, instrumental or mechanical procedures cannot be defined on the basis of their origin or of their essential properties. While ruler and compasses were arguably just the three-dimensional, concrete counterpart of straight line and circumference, sliding rulers, mesolabes *et similia* could not be easily reduced to geometrical objects, no matter how far one abstracted from their material qualities. The categories of "instrumental" and "mechanical" have no place in Pappus' classification, because they do not pin down any geometrical entity. Yet, when clarifying the distinction between planar, solid and linear problems, Pappus reminds the reader that the ancients were not able to find the two mean proportionals geometrically (τῷ γεωμετρικῷ λόγῳ), because conic sections were not easy to draw in a plane. Thus, he says, they managed to produce their constructions exclusively by means of instruments, because those latter were more expedient.[56] Mesolabes, moving or sliding rulers with pivots had no official place in the classification of problems, yet they were necessary for drawing both linear curves and, via a point-wise procedure, conic sections themselves.

Conic sections, then, were fully accepted in the domain of well-behaved geometrical objects, but Pappus is sometimes uneasy about their use; and as for mechanical procedures, although they do not enter the neat tripartite classification, they were practically indispensable. But, if its prescriptions are so easily skirted, what is the use of the classification after all? Pappus' partition of problems and methods seems to break down on questions not only and not mainly of consistency,

---

[56] 54.22–56.8. The same remarks, with only slight differences of phrasing, are found again at 1070.7–12, as an introduction to Pappus' own solution to cube-duplication.

but of significance for actual practice. As far as actual directions for problem-solving are concerned, it seems doomed to redundancy from the very start. At the moment of assigning a problem to one category rather than another, one must have at least an inkling of its solution already. That a problem is solid, for instance, is established not by any *a priori* characteristics that the problem may have: it cannot be decided with certainty from its enunciation, but must be an *a posteriori* consideration originating from its solution. For instance, in the case of establishing what category the duplication of the cube belonged to, Pappus must already have known a number of solid or linear solutions which worked, and perhaps some planar solutions which instead failed to determine the construction.

If the classification is indeed *a posteriori*, it would seem that, in the determination of what category a problem belongs to, the key-word is being aware of what others have done before you, legacy, tradition. Validity of procedure has to do not just with the characteristics of the problem, but also with the consensus established by the community, extended temporally to the "ancients" and laterally to the "new" mathematicians. The prescriptive function of Pappus' classification does more than stipulate what procedures should be used; it indicates what reference points and traditional knowledge an expert mathematician should be able to command. The rule will probably not signify at all for a reader who does not know what comes before.

Above all, imposing a classification is an empowering gesture: it shows that the author masters the entire domain covered by the classification, that he can put it in order and that he can draw lines between what is and is not allowed. Paradoxically, the very fact that Pappus does not always follow his own classification reinforces his role, in that he has command each time over the choice of what is the best criterion to follow. Imposing a classification is also a way of conveying a sense of bonding among experts – setting rules defines good practice, and spells out a shared set of procedures and results, dos and don'ts.

5.3 The ghost of mathematicians past

Pappus uses the mathematical tradition not just as a reservoir from which he can draw previous results, but also as a reference frame through which he can import meaning into the practice of classifying, for instance, or invest his own actions with authority by presenting himself as the true voice of the ancients. The *Collection* moves back and forth between past and present, and often contrasts earlier and more recent geometers.[57]

When introducing the classification of problems, Pappus notes that the ancients were at a loss to find a solution to the trisection of the angle; in the context of a similar passage, book 3 relates their difficulties with the duplication of the cube.[58] On a couple of occasions, Pappus even apologizes for the ancient geometers' bafflement: they were not well acquainted with conic sections, for instance, and that is why they made mistakes and failed to reach the results they wanted, in this case the trisection of the angle.[59] The moderns, on the other hand, have discovered how to apply conic sections to the trisection of the angle; they have proved the linear nature of the general division of arcs and angles; they have expanded the enquiry about linear curves; they have found a planar alternative to Archimedes' solid procedure based on the same premises.[60]

Pappus' attitude does not incline completely either way: sometimes newer geometers are attacked (e.g. the anonymous author in book 3) because of their ignorance or superficiality. In book 7, he laments the unnecessary multiplying of methodological distinctions, saying:

The ancients compiled their elements attending to the order of these plane *loci*; but the people who came after them disregarded this, and added others

---

[57] E.g. 270.2: οἱ παλαιοὶ γεωμέτραι; 272.8–9: οἱ πρότεροι γεωμέτραι; 270.24: οἱ νεώτεροι; 662.20: οἱ μετ' αὐτούς.

[58] 270.1–3 and 54.22 ff., respectively. On the history of the trisection of the angle see Hogendijk (1981).

[59] 270.1–3; 272.7–12.

[60] 272.12–14; 284.24; 270.24–25 and 302.16–18, respectively.

– as if they were not boundless in number if one wanted to add some that do not belong to that order! Hence I shall put the additional ones later, and those that belong to the order first, encompassing them by this one proposition.[61]

Then again, in book 6, he remarks: "Many of those who teach astronomical matters add some propositions as necessary, and leave aside others as unnecessary without paying them enough attention." Pappus mentions as examples a case where it has not been noted that a certain condition (that some circles cut each other perpendicularly) does not always hold; a case where it has not been specified in how many cases (ποσάκις) a certain thing holds; a third case where an author has been misinterpreted.[62] Further on in the same book, Pappus shows that one of the astronomical works (the proper selection of which he has just indicated as a task for the diligent mathematician) goes wrong on several counts. First of all, its author Theodosius (he of the *Spherics*) has omitted to examine a sub-case of a certain theorem; Pappus puts forward a lemma that enables him to consider the sub-case.[63] Then Pappus corrects a theorem by adding a certain condition to the enunciation, and also discusses the reasons why his own correction is better than others which had been proposed (we do not know by whom). One alternative correction is stupid (ἔστιν δὲ τοῦτο σφόδρα εὔηθες), because it does not realize that it is important to follow the order of argumentation; another gives something futile (εἰκαῖον) as its justification.[64] Pappus then provides the reasons why he himself makes the

---

[61] 662.19–24: οἱ μὲν οὖν ἀρχαῖοι ⟨εἰς τὴν⟩ τῶν ἐπιπέδων τούτων τόπων τάξιν ἀπο-βλέποντες ἐστοιχείωσαν, ἧς ἀμελήσαντες οἱ μετ᾽ αὐτοὺς προσέθη⟨καν ἑτέρους, ὡς οὐκ ἀπείρων τὸ πλῆθος ὄντων εἰ θέλοι τις προσγράφειν οὐ τῆς τάξεως ἐκείνης ἐχόμενα. θήσω οὖν τὰ μὲν προσκείμενα ὕστερα, τὰ δ᾽ ἐκ τῆς τάξεως πρότερα μιᾷ περιλαβὼν προτάσει ταύτη.

[62] 474.3–5: Πολλοὶ τῶν τὸν ἀστρονομούμενον τόπον διδασκόντων ἀμελέστερον τῶν προτάσεων ἀκούοντες τὰ μὲν προστιθέασιν ὡς ἀναγκαῖα, τὰ δὲ παραλείπουσιν ὡς οὐκ ἀναγκαῖα. The whole passage at 474.3–14.

[63] The lemma at 478.22–480.6; the sub-case (τοῦτο γὰρ οὐκ ἔδειξεν Θεοδόσιος) at 482.9–22. Another omission by Theodosius, which Pappus purports to prove "in a most astronomic way" (ἀστρονομικώτατα) at 532.6–536.7.

[64] The whole passage at 506.21–510.24; the quotations at 508.6 and 510.5 respectively.

correction, with a specific demonstration of how, if one does not add such-and-such a condition, the problem is not always solved. This depends on changing some elements in the diagram. After having examined four configurations, which correspond to as many sub-cases, Pappus sums up the results in a final proposition that takes into account all possible particularities.[65]

The same expertise that allows a geometer to make the right cuts across the material handed down to him by the tradition also entitles him to point out what in that tradition needs improving. Pappus is more than ready to magnify the achievements of an Archimedes, or the accomplishments of some bygone "all-round" scientists,[66] but he can also criticize Archimedes, or remark that "we do not have to trust the opinion [or the reputation] of the men who discovered [a certain thing]."[67]

Further evidence about Pappus' attitude to other geometers, this time Apollonius and Euclid, is contained in book 7:

The locus on three and four lines that [Apollonius] says, in [his account of] the third [book],[68] was not completed by Euclid, neither he nor anyone else would have been capable of; no, he could not have added the slightest thing to what was written by Euclid, at any rate using only the conics that had been proved up to Euclid's time, as he himself confesses when he says that it is impossible for it to be completed without what he himself was forced to establish first. But either Euclid, out of respect for Aristaeus as meritorious for the conics he had published already, did not anticipate him, or, because he did not desire to commit to writing the same matter as he [Aristaeus], – for he was the fairest of men, and kindly to everyone who was the slightest bit able to augment knowledge as one should, and he was not at all belligerent, and though exacting, not boastful, the way this man [Apollonius] was ... [Apollonius] was able to add the missing part to the locus because he had Euclid's writings on the locus already before him in his mind, and had studied for a long time in Alexandria under the people who had been taught

---

[65] The whole passage at 514.6–518.14. Pappus does something similar at 536.8–538.8: he criticizes someone and then provides the correct result.

[66] Book 8, 1022–1028, esp. 1026.5 ff.

[67] 254.23–24: οὐ χρὴ τῇ τῶν εὑρόντων ἀνδρῶν δόξῃ πιστεύοντας.

[68] Pappus' phrasing corresponds to our extant text of the introduction to the *Conics*, where Apollonius boasts of his novel discoveries, I 4.13–7.

by Euclid, where he also acquired such a great disposition [of mind], which was not without defect.[69]

Apollonius is rebuked not because he has got something wrong from a mathematical point of view, but for his attitude to some results of Euclid's. His behaviour is denounced in other, later, sources: Proclus, as we have seen in chapter 2, and Eutocius, who, quoting Heraclides, says that Apollonius tried to arrogate to himself discoveries made by Archimedes and not published for a larger public.[70] Pappus himself, while somehow trying to make up for Archimedes' mistake as denounced in book 4, does not do the same for Apollonius, who is after all accused of the same fallacy. It is difficult to imagine what evidence Pappus could be relying upon when he says that Euclid chose to write about the conics only to a limited extent out of respect for Aristaeus.[71] What is at stake here, however, is not the actual Euclid, Apollonius or Aristaeus: although they do represent themselves (as seen through Pappus' eyes), at the same time they are made to stand for something else.

The author of the *Elements* is presented as rich both in generic moral virtues and in mathematical ones – indeed, the

---

[69] 674.20–682.23: Ἀπολλώνιος μὲν ταῦτα. ὃν δέ φησιν ἐν τῷ τρίτωι τόπον ἐπὶ γ̄ καὶ δ̄ γραμμὰς μὴ τετελειῶσθαι ὑπὸ Εὐκλείδου, οὐδ' ἂν αὐτὸς ἐδυνήθη οὐδ' ἄλλος οὐδεὶς ἀλλ' οὐδὲ μικρόν τι προσθεῖναι τοῖς ὑπὸ Εὐκλείδου γραφεῖσιν διά γε μόνων τῶν προδεδειγμένων ἤδη κωνικῶν ἄχρι τῶν κατ' Εὐκλείδην, ὡς καὶ αὐτὸς μαρτυρεῖ λέγων ἀδύνατον εἶναι τελειωθῆναι χωρὶς ὧν αὐτὸς προγράφειν ἠναγκάσθη. ὁ δὲ Εὐκλείδης ἀποδεχόμενος τὸν Ἀρισταῖον ἀξιωθέντα ἐφ' οἷς ἤδη παρεδεδώκει κωνικοῖς καὶ μὴ φθάσας, ἢ μὴ θελήσας ἐπικαταβάλλεσθαι τούτῳ τὴν αὐτὴν πραγματείαν, ἐπιεικέστατος ὢν καὶ πρὸς ἅπαντας εὐμενὴς τοὺς καὶ κατὰ ποσὸν συνάξειν δυναμένους τὰ μαθήματα ὡς δεῖ καὶ μηδαμῶς προσκρουστικὸς ὑπάρχων, καὶ ἀκριβὴς μὲν οὐκ ἀλαζονικὸς δὲ καθάπερ οὗτος ... προσθεῖναι δὲ τῷ τόπῳ τὰ λειπόμενα δεδύνηται προφαντασιωθεὶς τοῖς ὑπὸ Εὐκλείδου γεγραμμένοις ἤδη περὶ τοῦ τόπου καὶ σχολάσας τοῖς [ὑπὸ] Εὐκλείδου μαθηταῖς ἐν Ἀλεξανδρείαι πλεῖστον χρόνον ὅθεν ἔσχεν καὶ τὴν τοσαύτην ἕξιν οὐκ ἀπαθῆ.

[70] Proclus, *In Eucl.* 113.3–19. Eutocius, *In Apol. Con.* 168.8 ff. cited in Knorr (1986a), 298. Cf. also *In Apol. Con.* 186.1–10, where Eutocius confirms that Apollonius criticized Euclid at the beginning of the *Conics*, but not, as Pappus and others have said, because Euclid had failed to inquire into the two mean proportionals. The bone of contention was instead a book on *loci* – I can only think that Eutocius' sources here are misunderstood.

[71] Cf. Knorr (1982b) and Jones's comments in (1986a), II 404.

two are never sharply separated. Euclid is fair and kind, not belligerent, not boastful – his two main virtues (which are emphasized by contrasting them with Apollonius' conduct, or by adding "as one should") are modesty and respect for other geometers, both contemporary and past: in other words, respect for the discipline itself, for the tradition.[72] Arrogance is condemned by Pappus on other occasions, most notably in his criticism of the anonymous geometer in book 3. One of the things that he is not forgiven is that he boasts that he has a solution when in fact he does not have one. This is more serious than stating the false; it also clouds falsity with an aura of assertiveness that makes it more difficult for other geometers to unmask the fallacy. In the case of Apollonius in book 7, modesty is seen once again as an *interactive* virtue: Aristaeus was a worthy mathematician, who had augmented knowledge and thus deserved Euclid's respect, which in its turn is presented as commendable behaviour. Arrogance, on the other hand, breaches mutuality because it makes a person ungrateful and inconsiderate. Apollonius managed to achieve results which, though not complete, were superior to Euclid's only because he could benefit from Euclid's writings and from other mathematicians in Alexandria who were well acquainted with those writings.[73] Apollonius' mistake, therefore, is chiefly that of not interrelating.

The two virtues depicted are thus two sides of the same coin, and they illustrate values that lie behind the desire to conjure up a community of geometers, whose aims, rules and deontology are agreed upon. Respect for the rules implies respect for the common aim (to augment knowledge) and for some basic principles that include recognition of authority and deference to one's spiritual fathers. Apollonius should have acknowledged his debts, and he did not. Euclid was not exempt from the rules himself just because he was a "greater" authority than Apollonius. The point is not that Euclid is

[72] Sachs (1917), 88–9 points out that some scholia to Euclid's *Elements* (V.654.1–10 to book 13 Introd.; V.282.13 to book 5 Introd.), which she takes to be by Pappus, have a "characteristic way of defending Euclid."
[73] 678.8–12.

more ancient, and consequently more worthy of respect, than Apollonius; nowhere is it said that Apollonius could *not* have improved on what Euclid had done. The point is rather that Apollonius should have recognized that he was walking along lines which had been drawn before him: he should have been aware of his history.

Conclusion

If we use "tradition" as a key-word to understand Pappus' mathematical practice, a lot of things fall into place. The features that we have identified as characteristic acquire significance; in some cases, in fact, they seem to have a meaning only if we read them in this light.

For instance, Pappus' classification assumes a knowledge of past results and procedures. Operations such as generalizations and analyses of particular cases presuppose in their turn the existence of previous, well established results, on the basis of which such manipulations can be carried out. Arithmetizations themselves can be seen as an exploration of new boundaries, an additional side to what it is to have mathematical knowledge of an object. They can be carried out as an application, an exercise or a determination of geometrical propositions which, again, are already known. In other words, Pappus' own ways of doing mathematics would not be possible without a mathematical tradition.

Saying this does not amount to repeating that Pappus was a compiler who stood in awe of his sources. Pappus' past is instrumental to his present. He selects his sources according to his audience, to the topic of the treatise, to whether he has anything to add to them or not. He quotes them selectively, modifying them according to his own interests, or declines to quote them, either because the reader should already be acquainted with them, or because the reader should not be expected to be acquainted with them, and can be told about the matter in Pappus' own words. The ancients can be criticized or praised, depending on the point Pappus wants to make. Our author is extremely present-orientated, and most of his moves must be read as attempts to augment his prestige, build an authoritative self-image, define an identity for his discipline and draw boundaries between mathematics and other forms of knowledge or, within mathematics itself, between good mathematicians and bad ones.

The fact that the *Collection* relies so heavily on the tradition, then, provides a way for its author to empower himself and make his practice persuasive and significant, but the tradition is tailored so as to serve precisely those aims. The mathematical tradition which we find in Pappus is a construct.

If we look again at the "outside world," we can see that what Pappus was doing was in a sense normal: tradition and the past are necessary keys to understanding most of the cultural activities of late antiquity. Commentaries, introductions, biographies were all common literary genres in fields such as philosophy, grammar, and medicine;[74] law became more and more officially constituted as a practice and a form of knowledge where tradition played a primary role. The so-called "Law of citations," issued in AD 426, ratified what had long been common practice when it established that some juridical sources, all dating back to the third century, should be the ultimate authority in controversial decisions.[75] Changes in the material aspects of the written word, namely the gradual passage from the roll to the codex, have also been seen as relevant to the role past works came to play in various branches of knowledge, in that the codex form facilitated reference, division into chapters and sections, marginal glossae, consultation, quotation – all practices that in a certain sense constitute and bring into existence a tradition.[76]

Finally, and once again in accord with what has been observed within other cultural practices at the time, the use of the tradition in Pappus is closely intertwined with his view of

[74] Cf. Stahl (1964); Manuli (1983); Donini (1982); Cox (1983); Giuffrida (1985); Kaster (1988); Geymonat (1990); Sharples (1990); Hadot (1991); Lizzi (1990); Gara (1992); Lloyd (1993); Romano (1994); Vegetti (1994). Cf. also, for an earlier period, Geffcken (1932); Fuhrmann (1960); Momigliano (1971).
[75] *Cod. Theod.* 1.4.3. Cf. also Cavallo (1975); Cambiano (1988b); Lim (1995).
[76] Cf. Kleberg (1962); Reynolds & Wilson (1968); Cavallo (1975), (1988), (1989a) and (1989b); Wilson (1984); Hopkins (1991); Pecere (1990); Canfora (1995).

mathematics as something defined not just by skill and knowledge, but also by a sort of ethical code, a set of shared values which define the mathematician as not just an expert, but also as a hero, to use Peter Brown's expressions.[77] We have seen this emerging on various occasions. When he takes issue with the philosophers, when he deals with mechanics and emphasizes its complementarity with mathematics, when he criticizes Apollonius or praises Euclid, Pappus is shaping an image of the mathematician as someone who is competent in a type of knowledge which is worthy of investigation for itself (it is useful and it fully realizes the qualities which make humans different from animals), and is at the same time the embodiment of ethical (political and social) qualities that were required of the members of the educated elite of his day.

The cultural life of the fourth century AD is obviously much more complex and problematic than anything I can sketch in a few sentences. My contention is simply that Pappus is a part of it and needs to be seen as such. His cultural context helps to explain Pappus, and he contributes to that context himself in showing how mathematics was a part of the picture.

[77] Cf. for instance Brown (1980).

BIBLIOGRAPHY

PRINCIPAL EDITION USED

Pappus Alexandrinus, *Collectio Mathematica*, ed. F. Hultsch, Berlin 1876–78

OTHER EDITIONS AND OTHER WORKS BY PAPPUS

Collectio Mathematica, Lat. trans. F. Commandino, Pisa 1588 (Pisa 1602, Bologna 1660)

Collectio Mathematica liber II, ed. J. Wallis, Oxford 1688

La collection mathématique, French trans. P. Ver Eecke, Paris/Bruges 1933

Book VII of the Collection, ed. and Engl. trans. A. Jones, New York/Berlin/Heidelberg/Tokyo 1986

In Ptolomaei Almagestum, ed. A. Rome, Città del Vaticano 1931

Commentary on book X of Euclid's Elements, Engl. trans. G. Junge & W. Thomson, Cambridge, Mass. 1930

PRIMARY SOURCES

Aelianus, *De Natura Animalium*, Engl. trans. A. F. Schofield, Cambridge, Mass./London 1971–72 (Loeb)

Aelius Aristides, *Hieros Logos*, Engl. trans. C. A. Behr, Leiden 1981

   Pro quattuor and *Scholia*, ed. W. Dindorf, Leipzig 1829

Agennius Urbicus, *Commentum ad Frontini De agrorum qualitate, De controversiis*, in *Corpus Agrimensorum Romanorum*, ed. K. Thulin, Leipzig 1913

   De controversiis agrorum, in *Corpus Agrimensorum Romanorum*, ed. K. Thulin, Leipzig 1913

Collection des anciens alchimistes grecs, ed. M. Berthelot & Ch.-Em. Ruelle, London 1963

Ammianus Marcellinus, *Rerum gestarum libri*, Engl. trans. J. C. Rolfe, Cambridge, Mass./London 1963 (Loeb)

Ammonius, *In Aristotelis Categorias*, Engl. trans. S. Marc Cohen & G. B. Matthews, London 1991

Aphthonius, *Progymnasmata*, ed. H. Rabe, Leipzig 1926

Apollonius, *Conica*, ed. J. L. Heiberg, Leipzig 1891–93

Conics. Books V to VII, ed. and Engl. trans. G. J. Toomer, New York 1990

Archimedes, *Opera*, ed. J. L. Heiberg, Leipzig 1910–15

Aristotle, *Analytica Posteriora*, Engl. trans. J. Barnes, Oxford 1975
> *De Anima*, Engl. trans. W. S. Hett, London 1957 (Loeb)
> *De Caelo*, Engl. trans. W. K. C. Guthrie, London 1939 (Loeb)
> *Physica*, Engl. trans. R. P. Hardie & R. K. Gaye, in J. Barnes (ed.), *The Complete Works of Aristotle*, Princeton 1984
> *Sophistici Elenchi*, Engl. trans. E. S. Forster, London/Cambridge, Mass. 1955 (Loeb)

[Aristotle], *Mechanica*, Engl. trans. W. S. Hett, London/Cambridge, Mass. 1963 (Loeb)

Athenaeus, *Deipnosophistae*, Engl. trans. C. Burton Gulick, London/New York 1928 (Loeb)

Augustine, *Confessiones*, ed. M. Skvtella, Stuttgart 1969 Engl. trans. H. Chadwick, Oxford 1991
> *De consensu evangelistarum*, ed. F. Weihrich, Vienna/Leipzig 1904
> *De ordine*, ed. W. M. Green, Turnhout 1970
> *Epistulae*, ed. A. Goldbacher, Vienna/Leipzig 1895–1923

Balbus, *Ad Celsum expositio et ratio omnium formarum*, in *Gromatici Veteres*, eds. F. Blume, K. Lachmann, A. Rudorff, Berlin 1848–52
> *Expositio et ratio mensurarum*, in *Gromatici Veteres*, eds. F. Blume, K. Lachmann, A. Rudorff, Berlin 1848–52

Biton, *Belopoietica* in Marsden (1971)

Cassiodorus, *Variae*, ed. T. Mommsen, Berlin 1894

Cicero, *De natura deorum*, Engl. trans. H. Rackham, Cambridge, Mass./London 1972 (Loeb)
> *De re publica*, Engl. trans. C. W. Keyes, Cambridge, Mass./London 1948 (Loeb)
> *Tusculanae disputationes*, Engl. trans. J. E. King, Cambridge, Mass./London 1950 (Loeb)

Claudianus, *Carmina minora*, Engl. trans. M. Platnauer, London/New York 1922 (Loeb)

Codex Justiniani, ed. E. Herrmann, 5th edn. Leipzig 1875

Codex Theodosiani, ed. T. Mommsen and P. M. Meyer, Berlin 1954

De isoperimetris figuris, ed. F. Hultsch, in Pappus Alexandrinus, *Collectio, cit.*

De rebus bellicis, ed. A. Giardina, Milan 1989; Engl. trans. E. A. Thompson, *A Roman Reformer and Inventor*, Oxford 1952

Descartes, R. (1637) *Géométrie*, Engl. trans. D. E. Smith & M. L. Latham, New York 1954

Digesta Justiniani, ed. T. Mommsen and P. Krueger; Engl. trans. A. Watson, Philadelphia 1985

Dio Cassius, *Historia*, Engl. trans. E. Cary, Cambridge, Mass./London 1969 (Loeb)

Diocles, *On Burning Mirrors*, ed. and Engl. trans. G. J. Toomer, Berlin/
Heidelberg/New York 1976
Diodorus Siculus, *Bibliotheca*, Engl. trans. C. H. Oldfather *et alii*, Cam-
bridge, Mass./London 1933–67 (Loeb)
Dorotheus Sidonius, *Carmen astrologicum*, ed. D. Pingree, Leipzig 1976
Euclid, *Data*, ed. H. Menge, Leipzig 1896
 Elementa, ed. J. L. Heiberg, with *Scholia*, Leipzig 1883–88
Eunapius, *Vitae philosopharum et sophistarum*, Engl. trans. W. Cave
Wright, Cambridge, Mass./London 1968 (Loeb)
Eusebius Caesarensis, *Historia ecclesiastica*, ed. & French trans. G. Bardy,
Paris 1955
Eutocius, *In Apollonii Conica Commentaria*, ed. J. L. Heiberg, in Apollo-
nius, *Opera*, Leipzig 1893
 Commentarium in Archimedis De sphaera et cylindro, ed. J. L. Heiberg, in
Archimedes, *Opera*, Leipzig 1915
 Commentarium In Archimedis Dimensionem Circuli, ed. J. L. Heiberg, in
Archimedes, *Opera*, Leipzig 1915
Expositio totius mundi, ed. J. Rougé, Paris 1966
Faventinus, *De diversis fabricis architectonicae*, ed. V. Rose, with Engl.
trans. by H. Plommer in H. Plommer, *Vitruvius and Later Building
Manuals*, Cambridge 1973
Julius Firmicus Maternus, *Matheseos Libri*, eds. W. Kroll, F. Skutsch & K.
Ziegler, Stuttgart 1968
Frontinus, *De agrorum qualitate, De controversiis, De limitibus, De arte
mensoria*, in *Corpus Agrimensorum Romanorum*, ed. K. Thulin, Leipzig
1913
 Stratagemata, ed. G. Gundermann, Leipzig 1888
Gregorius Nyssenus, *Epistulae*, ed. P. Maraval, Paris 1990
Hephaestio Thebanus, *Apotelesmatica*, ed. D. Pingree, Leipzig 1973
 Apotelesmaticorum epitomae, ed. D. Pingree, Leipzig 1974
Hero Alexandrinus, *Belopoeica*, Engl. trans. E. W. Marsden, in *Greek and
Roman Artillery: Technical Treatises*, Oxford 1971
 Definitiones, ed. J. L. Heiberg, Leipzig 1912
 Dioptra, ed. H. Schöne, Leipzig 1903
 Geometrica and *Scholia*, ed. J. L. Heiberg, Stuttgart 1976
 Mechanica, ed. L. Nix and W. Schmidt, Leipzig 1900
 Metrica, ed. H. Schöne, Leipzig 1903
Herodotus, *Historiae*, Engl. trans. A. D. Godley, Cambridge, Mass./
London 1926 (Loeb)
Hieronymus, *Epistulae*, ed. J. Labourt, Paris 1953
Hippolytus, *Refutatio omnium haeresium*, ed. M. Marcovich, Berlin/New
York 1986
Homer, *Iliad*, Engl. trans. A. T. Murray, Cambridge, Mass./London 1978
(Loeb)

Hyginus Gromaticus, *De limitibus constituendis*, in *Corpus Agrimensorum Romanorum*, ed. K. Thulin, Leipzig 1913

Hypsicles, *Elementorum Liber XIV*, in Euclid, *Elementa*, ed. J. L. Heiberg, Leipzig 1888

Iamblichus, *Adhortatio ad philosophiam*, Engl. trans. T. Moore Johnson, Grand Rapids 1988

    In Nicomachi Arithmeticam Introductionem, ed. E. Pistelli, Leipzig 1894

[Iamblichus], *Theologia arithmetica*, Engl. trans. R. Waterfield, Grand Rapids 1988

Inscriptiones Latinae Selectae, ed. H. Dessau, Berlin 1906

Isocrates, *Antidosis*, Engl. trans. G. Norlin, London/New York 1929 (Loeb)

    Panathenaicus, Engl. trans. G. Norlin, London/New York 1929 (Loeb)

Julian, *Contra Galileas*, Engl. trans. W. C. Wright, London 1913–23

Lactantius Firmianus, *Divinae Institutiones*, ed. P. Morrat, Paris 1973–87

Libanius, *Epistulae*, ed. R. Foerster, Leipzig 1903–27

Johannes Lydus, *De magistratibus romanis*, ed. A. C. Bandy, Philadelphia 1983

Mamertinus, *Gratiarum actio divi Juliani Augusti*, in Nixon & Saylor Rodgers (1994)

Manilius, *Astronomica*, ed. with Engl. trans. G. P. Good, Cambridge, Mass./London 1992 (Loeb)

Marinus, *In Euclidis Data*, in Euclid, *Opera*, ed. J. L. Heiberg & H. Menge, Leipzig 1896

Marcus Valerius Martialis, *Epigrammata*, ed. with Engl trans. D. R. Shackleton Bailey, Cambridge, Mass./London 1993 (Loeb)

Martianus Capella, *De nuptiis Philologiae et Mercurii*, ed. J. Wallis, Leipzig 1983

Marcus Junius Nipsus, *Fluminis varatio, Limitis repositio, Podismus*, in *Gromatici Veteres*, eds. F. Blume, K. Lachmann, A. Rudorff, Berlin 1848–52

Notitia Dignitatum, ed. O. Seeck, 1876, reprinted Frankfurt a. M. 1962

Origen, *Contra Celsum*, ed. M. Borret, Paris 1969

Ovidius, *Fasti*, Engl. trans. J. G. Frazer, Cambridge, Mass./London 1959 (Loeb)

Paulus Alexandrinus, *Elementa Apotelesmatica*, ed. E. Boer, Leipzig 1958

Philo Byzantinus, *Belopoeica*, Engl. trans. E. W. Marsden, in *Greek and Roman Artillery: Technical Treatises*, Oxford 1971

Philostratus, *Vitae Sophistarum*, Engl. trans. W. Cave Wright, London/ New York 1922

Photius, *Bibliotheca*, ed. O. Bekker, Berlin 1824

Plato, *Leges*, ed. J. Burnet, Oxford 1906

    Phaedo, Engl. trans. H. N. Fowler, Cambridge, Mass./London 1960 (Loeb)

Res Publica, Engl. trans. P. Shorey, Cambridge, Mass./London 1946 (Loeb)

Timaeus, Engl. trans. R. G. Bury, Cambridge, Mass./London 1966 (Loeb)

Plinius Secundus, *Epistulae*, Engl. trans. W. Melmoth & W. M. L. Hutchinson, Cambridge, Mass./London 1947 (Loeb)

Plotinus, *Enneades*, Engl. trans. A. H. Armstrong, Cambridge, Mass./London 1988 (Loeb)

Plutarch, *Moralia: Apophthegmata Laconica*, Engl. trans. F. Cole Babbitt, Cambridge, Mass./London 1931 (Loeb)

Moralia: De E apud Delphos, Engl. trans. F. C. Babbitt, Cambridge, Mass./London 1969 (Loeb)

Moralia: De Genio Socratis, Engl. trans. P. H. De Lacy & B. Einarson, Cambridge, Mass./London 1968 (Loeb)

Moralia: Quaestiones Conviviales, Engl. trans. E. L. Minar, Cambridge, Mass./London 1961 (Loeb)

Moralia: regum et imperatorum apophthegmata, Engl. trans. F. Cole Babbitt, Cambridge, Mass./London 1931 (Loeb)

Vitae Parallelae: Marcellus-Pelopidas, Engl. trans. B. Perrin, Cambridge, Mass./London 1968 (Loeb)

Vitae Parallelae: Demetrius-Antonius, Engl. trans. B. Perrin, London/New York 1920 (Loeb)

Polybius, *Historiae*, Engl. trans. W. R. Paton, Cambridge, Mass./London 1960 (Loeb)

Proclus, *In Platonis Timaeum commentaria*, ed. E. Diehl, Leipzig 1903–6

In Primum Euclidis Elementorum Librum Commentaria, ed. G. Friedlein, Leipzig 1873

Procopius, *De aedificiis libri*, ed. J. Haury, Leipzig 1964

De bellis libri, ed. G. Wirth, Leipzig 1962

Ptolemy, *Apotelesmatica (Tetrabiblos)*, ed. F. Boll & A. Boer, Leipzig 1957; Engl. trans. G. P. Goold, Cambridge, Mass./London 1977 (Loeb)

Planetary Hypotheses, ed. J. L. Heiberg & L. Nix, Leipzig 1907

Syntaxis Mathematica (Almagest), ed. J. L. Heiberg, Leipzig 1898

Marcus Fabius Quintilianus, *Institutio Oratoria*, Engl. trans. H. E. Butler, Cambridge, Mass./London 1963 (Loeb)

Scriptores Historiae Augustae, ed. E. Hohl, Leipzig 1927

Siculus Flaccus, *De condicionibus agrorum*, in *Corpus Agrimensorum Romanorum*, ed. K. Thulin, Leipzig 1913

De condicionibus agrorum, ed. & French trans. M. Clavel-Lévêque, D. Conso, F. Favory, J.-Y. Guillaumin, P. Robin, Napoli 1993

Silius Italicus, *Punica*, Engl. trans. J. D. Duff, Cambridge, Mass./London 1950 (Loeb)

Simplicius, *In Aristotelis Categorias Commentaria*, ed. K. Kalbfleisch, Berlin 1907

In Aristotelis De Caelo Commentaria, ed. J. L. Heiberg, Berlin 1894
In Aristotelis Physicam Commentaria, ed. H. Diels, Berlin 1882–95
Stobaeus Johannes, *Anthologia*, ed. A. Meineke, Leipzig 1857
Suidae Lexicon, ed. A. Adler, Leipzig 1928–38
Suetonius, *Opera*, Engl. trans. J. C. Rolfe, Cambridge, Mass./London 1914
    (Loeb)
Symmachus, *Epistulae*, ed. and French trans. J.-P. Callu, Paris 1982
    Relationes, ed. D. Vera, Pisa 1981
Theon Alexandrinus, *In Ptolomaei Almagestum*, ed. A. Rome, Vatican 1936
Theon of Smyrna, *Expositio Rerum Mathematicarum ad Legendum Plato-
    nem Utilium*, ed. E. Hiller, Leipzig 1878
Johannes Tzetzes, *Commentarium in Aristophanis Nubes*, ed. D. Holwerda,
    Groningen/Amsterdam 1960
    Chiliades, ed. P. A. Leone, Napoli 1968
Ulpianus, *De excusatione*, in *Iurisprudentiae anteiustinianae reliquiae*, eds.
    P. E. Huschke, E. Seckel, B. Kuebler, Leipzig 1911
Valentinianus, *Novellae*, in *Codex Theodosiani*
Vegetius, *Epitoma rei militaris*, Engl. trans. L. F. Stelten, New York/Bern/
    Frankfurt a.M./Paris 1990
Vergil, *Aeneid*, ed. W. Ianell, Leipzig 1930
Vettius Valens, *Anthologiae*, ed. D. Pingree, Leipzig 1986
Vitruvius, *De architectura*, Engl. trans. F. Granger, Cambridge, Mass./
    London 1962 (Loeb)
Die Fragmente der Vorsokratiker, eds. H. Diels & W. Kranz Dublin/Zurich
    1966
Zenodorus, *De isoperimetris figuris*, in Theon Alexandrinus, *In Almagestum*,
    ed. A. Rome, Vatican 1936

SECONDARY SOURCES

Anderson, J. C. Jr. (1997) *Roman Architecture and Society*, Baltimore/
    London
Anton, J. P. (1980) *Science and the Sciences in Plato*, New York
Armstrong, A. H. (1984) "Pagan and Christian traditionalism in the first
    three centuries AD," *Studia patristica* 15: 414–431
Aujac, G. (1970) "La sphéropée, ou la mécanique au service de la décou-
    verte du monde," *Revue d'histoire des sciences* 23: 93–107
    (1984) "La Langue formulaire dans la géométrie grecque " *Revue d' his-
    toire des sciences et leurs applications* 37.2: 97–109
Authier, M. (1989) "Archimedes: the scientist's canon," in M. Serres (ed.) *A
    History of Scientific Thought* (Oxford/Cambridge, Mass. 1995) 124–
    159, Engl. trans. of *Eléments d' histoire des sciences*, Paris
Baccani, D. (1992) *Oroscopi greci. Documentazione papirologica*, Messina
Bagnall, R. S. (1993) *Egypt in Late Antiquity*, Princeton
Baldwin, B. (1978) "The De Rebus Bellicis," *Eirene* 16: 23–29

(1982) "Literature and society in the late Roman Empire," in B. K. Gold (ed.) *Literary and Artistic Patronage in Ancient Rome*, (Austin) 67–83

Barton, T. S. (1994a) *Ancient Astrology*, London/New York

(1994b) *Power and Knowledge. Astrology, Physiognomics and Medicine under the Roman Empire*, Ann Arbor

Beavis, I. C. (1988) *Insects and Other Invertebrates in Classical Antiquity*, Exeter

Behboud, A. (1994) "Greek geometrical analysis," *Centaurus* 37: 52–86

Berenson, B. (1954) *The Arch of Constantine or The Decline of Form*, London

Beretta, G. (1994) *Ipazia di Alessandria*, Rome

Bertelli, C. (ed.) (1988) *Il mosaico*, Milan

Berve, H. (1959) "König Hieron II," *Abhandlungen der Bayerischen Akademie der Wissenschaften*, phil.-hist. Kl. N.F.H. 47

Betz, H. D. (1982) "The formation of authoritative tradition in the Greek magical papyri," in B. F. Meyer and E. P. Sanders (eds.) *Jewish and Christian Self-definition in the Graeco-Roman World* (London) 161–170

Biagioli, M. (1989) "The social status of Italian mathematicians, 1450–1600," *History of Science* 27: 41–95

Böker, R. (1961) "Würfelverdoppelung," in *PW*, ser. II, 9, cols. 1193–1223 (1962), "Νεῦσις," in *PW* Supplement IX, cols. 415–461

Bos, H. (1996) "Tradition and modernity in early modern mathematics: Viète, Descartes and Fermat," in C. Goldstein, J. Gray and J. Ritter (eds.) *L'Europe mathématique. Histoires, mythes, identités* (Paris) 183–204

Boulvert, G. (1970) *Esclaves et affranchis impériaux sous le Haut-Empire romain. Rôle politique et administratif*, Naples

Boyaval, B. (1977) "Sur quelques exercices d'arithmétique et de géometrie," *Chronique d'Egypte* 52: 311–315

Boyer, C. B. (1939) *The Concept of the Calculus*, New York:

Bowen, A. C. (1983) "Menaechmus versus the Platonists: two theories of science in the early Academy," *Ancient Philosophy* 3: 12–29

Bowersock, G. W. (1969) *Greek Sophists in the Roman Empire*, Oxford (1990) *Hellenism in Late Antiquity*, Cambridge

Bowman, A. K. (1976) "Papyri and Roman imperial history, 1960–75," *The Journal of Roman Studies* 66: 153–173

(1980) "The economy of Egypt in the earlier fourth century," in C. E. King (ed.) *Imperial Revenue, Expenditure and Monetary Policy in the Fourth Century AD* (Oxford) 23–40

(1986) *Egypt after the Pharaohs. 332 BC–AD 642 from Alexander to the Arab Conquest*, London

(1992) "Public buildings in Roman Egypt," review of A. Lukaszewicz, *Les édifices publiques dans les villes de l'Egypte romaine: problèmes administratifs et financiers*, Warsaw 1986, in *Journal of Roman Archaeology* 5: 495–503

Brown, M. (1975) "Pappus, Plato and the harmonic mean,' *Phronesis* 20: 173–184

Brown, P. (1978) *The Making of Late Antiquity*, Cambridge, Mass.

(1980) "The philosopher and society in late antiquity," in E. C. Hobbs and W. Wuellner (eds.), *Protocol of the 34th Colloquy, Center for Hermeneutical Studies in Hellenistic and Modern Culture 1978* (Berkeley, Calif.) 1–41

(1992) *Power and Persuasion in Late Antiquity: Towards a Christian Empire*, Madison

Brunt, P. A. (1975) "The administration of Roman Egypt," *The Journal of Roman Studies* 64: 124–47

(1983) "Princeps and equites," *The Journal of Roman Studies* 73: 42–75

Bulmer-Thomas, I. (1974) "Pappus," *DSB*

(ed.) (1967) *Greek Mathematical Works*, Cambridge, Mass/London (Loeb)

(1981) review of J. Warren, *Greek Mathematics and the Architects to Justinian*, London 1976, *Historia Mathematica* 8: 482–435

Burkert, W. (1962) *Weisheit und Wissenschaft: Studien zu Pythagoras, Philolaos und Platon*, Nürnberg; revised Engl. trans. E. L. Minar Jr., *Lore and Science in Ancient Pythagoreanism*, Cambridge, Mass. 1972

Busard, H. L. L. (1980) "Der Traktat *de Isoperimetris*," *Mediaeval Studies* 42: 61–88

Calderini, A. (1920) "Appunti di terminologia secondo i documenti dei papiri," *Aegyptus* 1: 309–317

Cambiano, G. (1985a) "Figura e numero," in M. Vegetti (ed.), *Il sapere degli antichi* (Turin) 83–108

(1985b) "Proclo e il libro di Euclide," in C. Giuffrida and M. Mazza (eds.), *Le trasformazioni della cultura nella tarda antichità* (Roma) 265–279

(1988a) "La scrittura della dimostrazione in Grecia," in *Le savoir de l'écriture en Grèce ancienne* (Lille) 251–272

(1988b) "Sapere e testualità nel mondo antico," in P. Rossi (ed.) *La memoria del sapere. Forme di conservazione e strutture organizzative dall'antichità ad oggi* (Rome/Bari) 69–98

Cameron, Alan (1965) "Roman school fees," *The Classical Review* 15: 257–258

(1984) "The Latin revival of the fourth century," in W. Treadgold (ed.), *Renaissances before the Renaissance. Cultural Revival of Late Antiquity and the Middle Ages* (Stanford) 42–58

(1990) "Isidore of Miletus and Hypatia: on the editing of mathematical texts," *Greek, Roman and Byzantine Studies* 31: 103–127

Cameron, Averil (1993) *The Later Roman Empire AD 284–430*, London

Campbell, B. (1996) "Shaping the rural environment: surveyors in ancient Rome," *The Journal of Roman Studies* 86: 74–99

Canfora, L. (1986) *La biblioteca scomparsa*, Palermo

(1995) "Libri e biblioteche," in G. Cambiano, L. Canfora, D. Lanza (eds.), *Lo spazio letterario di Roma antica* (Roma) 11–93

Cantor, M. (1907) *Vorlesungen über Geschichte der Mathematik*, Leipzig

Cauderlier, P. (1978) "Sciences pures et sciences appliquées dans l'Egypte romaine. Essai d'inventaire antinoite," in J.-M. André (ed.), *Recherches sur les artes à Rome* (Paris) 47–76

Cavallo, G. (1975) "Libro e pubblico alla fine del mondo antico," in G. Cavallo (ed.), *Libri, editori e pubblico nel mondo antico* (Rome/Bari) 81–132

(1984) "Scuola, scriptorium, biblioteca a Cesarea," in G. Cavallo (ed.), *Le biblioteche nel mondo antico e medievale* (Rome/Bari) 65–79

(1988) "Cultura scritta e conservazione del sapere: dalla Grecia antica all'occidente medievale," in P. Rossi (ed.) *La memoria del sapere. Forme di conservazione e strutture organizzative dall'antichità ad oggi* (Rome/Bari) 29–67

(1989a) "Libro e cultura scritta," in A. Schiavone (ed.), *Storia di Roma. Caratteri e morfologie* (Turin) 693–734

(1989b) "Testo, libro e lettura," in G. Cavallo, P. Fedeli and A. Giardina (eds.), *Lo spazio letterario di Roma antica* (Rome) 307–341

Chadwick, H. (1947) "Origen, Celsus, and the Stoa," *The Journal of Theological Studies* 48: 34–49

Chihara, C. S. (1990) *Constructibility and Mathematical Existence*, Oxford

Chouquer, G. & Favory, F. (1992) *Les arpenteurs romains. Théorie et pratique*, Paris

Clagett, M. (1956) *Greek Science in Antiquity*, New York
(1964–84) *Archimedes in the Middle Ages* (5 vols.), Madison

Clarke, M. L. (1971) *Higher Education in the Ancient World*, London

Clavel-Lévêque, M. (1992) "Centuriation, géométrie et harmonie. Le cas du Biterrois," in J.-Y. Guillaumin (ed.) *Mathématiques dans l'antiquité* (Saint-Étienne) 161–184

Cook, A. B. (1895) "The bee in Greek mythology," *The Journal of Hellenic Studies* 15: 1–24

Corcoran, S. (1996) *The Empire of the Tetrarchs. Imperial Pronouncements and Government AD 284–324*, Oxford

Cornford, F. M. (1932) "Mathematics and dialectic in the Republic VI–VII," *Mind* 41: 37–52; 173–190

Coulton, J. J. (1977) *Greek Architects at Work. Problems of Structure and Design*, London

Cox, P. (1983) *Biography in Late Antiquity – A Quest for the Holy Man*, Berkeley/Los Angeles/London

Cracco Ruggini, L. (1971) "Le associazioni professionali nel mondo romano-bizantino," in *Artigianato e tecnica nella società dell'alto medioevo occidentale* (Spoleto) 59–227

(1985) "Arcaismo e conservatorismo, innovazione e rinnovamento (IV–V secolo)," in C. Giuffrida and M. Mazza (eds.), *Le trasformazioni della cultura nella tarda antichità* (Rome) 133–156

(1993) "Scienze pure e scienze applicate nella cultura tardo antica," in A. Schiavone (ed.) *Storia di Roma* (Turin) 839–863

Cumont, F. (1937) *L'Egypte des astrologues*, Brussels

Cuomo, S. (forthcoming) "The machine and the city: Hero of Alexandria's belopoietics," in C. Tuplin (ed.) *Science Matters* (Oxford)

Davies, M. & Kathirithamby, J. (1986) *Greek Insects*, London

De Gandt, F. (1982) "Force et science des machines," in J. Barnes *et al.* (eds.) *Science and Speculation. Studies in Hellenistic Theory and Practice* (Cambridge/Paris) 96–127

Delmaire, R. (1989a) *Les responsables des finances impériales au Bas-Empire romain (IVe–VIe): études prosopographiques*, Brussels

  (1989b) *Largesses sacrées et res privata: l'aerarium impérial et son administration du IVe au VIe siècle*, Rome

Detienne, M. & Vernant, J.-P. (1974) *Les Ruses d'Intelligence*, Paris, Engl. trans. by J. Lloyd as *Cunning Intelligence in Greek Culture and Society*, 1978

Dicks, D. R. (1972) "Geminus," *DSB*

Diels, H. (1924) *Antike Technik*, Leipzig/Berlin

Dilke, O. A. W. (1971) *The Roman Land Surveyors. An Introduction to the Agrimensores*, Newton Abbot

Dillon, J. M. (1982) "Self-definition in later Platonism," in B. F. Meyer and E. P. Sanders (eds.), *Jewish and Christian Self-Definition. Self-Definition in the Graeco-Roman World* (London) 60–75

Donderer, M. (1996) *Die Architekten der späten Römischen Republik und der Kaiserzeit. Epigraphische Zeugnisse*, Erlangen

Donini, P. (1982) *Le scuole, l'anima, l'Impero: la filosofia antica da Antioco a Plotino*, Turin

Downey, G. (1946–48) "Byzantine architects: their training and methods," *Byzantion* 18: 99–118

  (1947–48) "Pappus of Alexandria on architectural studies," *Isis* 38: 197–200

Drabkin, I. E. & Drake, S. (1969) *Mechanics in Sixteenth-Century Italy*, Madison

Dupont, C. (1967) "Les privilèges des clercs sous Constantin," *Revue d'histoire ecclésiastique* 62: 729–752

Dzielska, M. (1995) *Hypatia of Alexandria*, Engl. trans F. Lyra, Cambridge, Mass./London

Edwards, C. H. (1979) *The Historical Development of the Calculus*, New York/Heidelberg/Berlin

Elliott, T. G. (1978) "The tax exemptions granted to clerics by Constantine and Constantius II," *Phoenix* 32: 326–336

Étienne, E. & Roels, J. (1986) "Deux aspects particuliers du problème des moyennes dans Pappus d'Alexandrie," *Revue des questions scientifiques* 157: 179–198

Eves, H. (1976) *An Introduction to the History of Mathematics*, New York

Évrard, É. (1977) "A quel titre Hypatie enseigna-t-elle la philosophie?" *Revue des études grecques* 90: 69–74

Farrington, B. (1946) *Science and Politics in the Ancient World*, London

Fedeli P. (1989) "I sistemi di produzione e diffusione," in G. Cavallo, P. Fedeli and A. Giardina (eds.) *Lo spazio letterario di Roma antica* (Roma) 343–378

Ferrari, G. A. (1984) "Meccanica allargata," *Atti del Convegno 'La scienza ellenistica' Pavia 1982* (Naples) 227–296

    (1985) "Macchina e artificio," in M. Vegetti (ed.), *Il sapere degli antichi* (Turin) 163–179

Finley, M. (1965) "Technical innovation and economic progress in the ancient world," *The Economic History Review* 18: 29–45

Fleury P. (1994) "Héron d'Alexandrie et Vitruve. A propos des techniques dites 'pneumatiques' ", in G. Argoud (ed.) *Science et vie intellectuelle à Alexandrie (Ier–IIIe siècle après J.-C.)* (Saint-Étienne) 67–81

    (1996) "Traités de mécanique et textes sur les machines," in C. Nicolet (ed.) *Les littératures techniques dans l'antiquité romaine. Statut, public et destination, tradition* (Geneva) 45–75

Folkerts, M. (1992) "Mathematische Probleme im Corpus Agrimensorum," in O. Behrends and L. Capogrossi Colognesi (eds.) *Die Römische Feldmeßkunst. Interdisziplinäre Beiträge zu ihrer Bedeutung für die Zivilisationsgeschichte Roms* (Göttingen) 311–336

Forbes, R. J. (1960) *Man the Maker: A History of Technology and Engineering*, New York

Forlin Patrucco, M. (1985) "Forme della tradizione nella grecità tarda: la citazione classica come linguaggio politico," in C. Giuffrida and M. Mazza (eds.) *Le trasformazioni della cultura nella tarda antichità* (Rome) 185–203

Fowden, G. (1977) "The Platonist philosopher and his circle in late antiquity," Φιλοσοφια 7: 359–383

    (1982) "The pagan holy man in late antique society," *The Journal of Hellenic Studies* 102: 33–59

Fowler, D. H. (1987) *The Mathematics of Plato's Academy: A New Reconstruction*, Oxford

Franco Repellini, F. (1989) "Tecnologie e macchine," in A. Schiavone (ed.) *Storia di Roma* (Turin) 323–368

Frank, T. (1940) *An Economic Survey of Ancient Rome*, Baltimore

Fraser, P. M. (1972) *Ptolemaic Alexandria*, Oxford

Fuhrmann, M. (1960) *Das systematische Lehrbuch. Ein Beitrag zur Geschichte der Wissenschaften in der Antike*, Göttingen

Gabba, E. (ed.) (1984) "Scienza e potere nel mondo ellenistico," in *Atti del Convegno 'La scienza ellenistica' Pavia 1982* (Naples) 13–37

Gara, A. (1992) "Progresso tecnico e mentalità classicista," in A. Schiavone (ed.) *Storia di Roma* (Turin) II.3: 361–380

(1994) *Tecnica e tecnologia nelle società antiche*, Rome

Geffcken, J. (1932) "Zur Entstehung und zum Wesen des Griechischen Wissenschaftlichen Kommentars," *Hermes* 67: 397–412

Gericke, H. (1992) *Mathematik in Antike und Orient*, Wiesbaden

Geymonat, M. (1990) "Le mediazioni," in G. Cavallo, P. Fedeli and A. Giardina (eds.) *Lo spazio letterario di Roma antica* (Rome) 279–295

Gille, B. (1980) *Les mécaniciens grecs*, Paris

Ginzburg, C. (1983) "Clues: Morelli, Freud and Sherlock Holmes," in U. Eco and T. A. Sebeok (eds.) *The Sign of Three* (Bloomington) 81–111

Giuffrida, C. (1985) "*Disciplina Romanorum*. Dall'*Epitoma* di Vegezio allo ΣΤΡΑΤΗΓΙΚΟΝ dello pseudo-Mauricius," in C. Giuffrida and M. Mazza (eds.) *Le trasformazioni della cultura nella tarda antichità* (Rome) 837–860

Goguey, D. (1978) "La formation de l'architecte: culture et technique," in J.-M. André (ed.) *Recherches sur les artes à Rome* (Paris) 100–115

Goldstein, C. (1989) "Stories of the circle," in M. Serres (ed.) *A History of Scientific Thought* (Oxford/Cambridge, Mass. 1995) 160–190, Engl. trans. of *Eléments d'histoire des sciences*, Paris

(1996) "À la recherche des origines: contenus, sources, communautés et histoires," in C. Goldstein, J. Gray and J. Ritter (eds.) *L'Europe mathématique. Histoires, mythes, identités* (Paris) 15–30

Goldstein, C., Gray, J. and Ritter, J. (1996) "Introduction," in *iidem* (eds.) *L'Europe mathématique. Histoires, mythes, identités* (Paris) 7–12

Grégoire, H. (1927–8) "Inscriptions historiques byzantines," *Byzantion* 4: 437–468

Griffin, M. (1971) review of G. W. Bowersock, *Greek Sophists in the Roman Empire*, Oxford 1969, *The Journal of Roman Studies* 61: 278–280

Grodzynski, D. (1974) "Par la bouche de l'empereur," in J.-P. Vernant *et al.*, *Divination et rationalité* (Paris) 267–294

Guillaumin, J.-Y. (1988) "Les différents noms de l'angle chez les agrimensores latins," *Revue des études anciennes* 90: 411–417

(1992) "La signification des termes *contemplatio* et *observatio* chez Balbus et l'influence héronienne sur le traité," in J.-Y. Guillaumin (ed.) *Mathématiques dans l'antiquité* (Saint-Étienne) 205–214

Gulley, N. (1958) "Greek geometrical analysis," *Phronesis* 3: 1–14

Gyekye, K. (1972) "Al-Farabi on analysis and synthesis," *Apeiron* 6: 33–38

Haas, C. J. (1997) *Alexandria in Late Antiquity. Topography and Social Conflict*, Baltimore/London

Hadot, I. (1984) *Arts libéraux et philosophie dans la pensée antique*, Paris

(1991) "The role of the commentaries on Aristotle in the teaching of philosophy according to the prefaces of the Neoplatonic commentaries on the *Categories*," in H. Blumenthal & H. Robinson (eds.) *Aristotle and the Later Tradition* (Oxford) 175–189

Harvey, F. D. (1965) "Two kinds of equality," *Classica et Mediaevalia* 26: 101–146

Heath, T. L. (1921) *A History of Greek Mathematics*, Oxford
(1956) *The Thirteen Books of Euclid's Elements*, New York

Hewsen, R. H. (1971) "The Geography of Pappus of Alexandria. A translation of the Armenian fragments," *Isis* 62: 186–207

Hinrichs, F. T. (1974) *Die Geschichte der gromatischen Institutionen*, Wiesbaden
(1992) "Die 'agri per extremitatem mensura comprehensi.' Diskussion eines Frontinstextes und der Geschichte seines Verständnisses" in O. Behrends and L. Capogrossi Colognesi (eds.) *Die Römische Feldmeßkunst. Interdisziplinäre Beiträge zu ihrer Bedeutung für die Zivilizationsgeschichte Roms* (Göttingen) 348–374

Hintikka, J. & Remes, U. (1974) *The Method of Analysis: Its Geometrical Origin and its General Significance*, Dordrecht

Hjelmslev, J. (1950) "Eudoxus' axiom and Archimedes' lemma," *Centaurus* 1: 2–11

Hogendijk, J. P. (1981) "How trisections of the angle were transmitted from Greek to Islamic geometry," *Historia Mathematica* 8: 417–438

Hopkins, K. (1980) "Taxes and trade in the Roman Empire (200 BC–AD 400)," *The Journal of Roman Studies* 70: 101–125
(1991) "Conquest by book," in *Literacy in the Roman World*, special issue of *Journal of Roman Archaeology* 3: 133–158

Houston, G. W. (1989-90) "The state of the art: current work in the technology of ancient Rome," *The Classical Journal* 85: 63–80

Hφyrup, J. (1996a) "Changing trends in the historiography of Mesopotamian mathematics: An insider's view," *History of Science* 34: 1–32
(1996b) "The formation of a myth: Greek mathematics – our mathematics," in C. Goldstein, J. Gray and J. Ritter (eds.) *L'Europe mathématique. Histoires, mythes, identités* (Paris) 102–119
(1996c) "Hero, ps.-Hero, and Near Eastern practical geometry. An investigation of *Metrica, Geometrica*, and other treatises," in T. Berggren (ed.) *Third International Conference on Ancient Mathematics. Proceedings of the Third Biennial Meeting* (Burnaby) 96–116

Hübner, W. (1990) *Die Begriffe "Astrologie" und "Astronomie" in der Antike. Wortgeschichte und Wissenschaftssystematik, mit einer Hypothese zum Terminus "Quadrivium"*, Stuttgart

Ireland, R. (ed.) (1979) *De rebus bellicis*, Engl. trans. and commentaries, Oxford

Jackson, D. E. P. (1970) *The Arabic Version of the Mathematical Collection of Pappus Alexandrinus Book VIII*, unpublished PhD dissertation, University of Cambridge
(1972) "The Arabic translation of a Greek manual of mechanics," *The Islamic Quarterly* 16: 96–103

(1980) "Towards a resolution of the problem of τὰ ἑνὶ διαστήματι γρα-φόμενα in M. Pappus' collection Book VIII," *Classiccl Quarterly* 30: 523–528

Jones, A. (1986a) see Pappus Alexandrinus, *Book VII of the Collection*

(1986b) "William of Moerbeke, the papal Greek manuscripts, and the Collection of Pappus of Alexandria in Vat. gr. 218," *Scriptorium* 40: 16–31

(1994a) "The place of astronomy in Roman Egypt," in T. D. Barnes (ed.) *The Sciences in Greco-Roman Society*, special issue of *Apeiron* 27: 25–51

(1994b) "Later Greek and Byzantine mathematics," in I. Grattan-Guinness (ed.) *Companion Encyclopedia of the History and Philosophy of the Mathematical Sciences* (London/New York) 1: 64–69

Jones, A. H. M. (1964) *The Later Roman Empire 284–602. A Social, Economic, and Administrative Survey*, Oxford

Karpinski, L.-C. (1923) "Michigan mathematical papyrus r.o. 621," *Isis* 5: 20–25

Kaster, R. A. (1983) "Notes on 'primary' and 'secondary' schools in late antiquity," *Transactions of the American Philological Association* 113: 323–346

(1988) *Guardians of Language: The Grammarian and Society in Late Antiquity*, Berkeley/Los Angeles/London

Kayser, F. (1994) *Recueil des inscriptions grecques et latines (non funéraires) d'Alexandrie impériale (i^er-IIIe s. apr. J.-C.)*, Cairo

Kelly, C. (1994) "Later Roman bureaucracy: going through the files," in A. K. Bowman and G. Woolf (eds.) *Literacy and Power in the Ancient World* (Cambridge) 161–176

(1998) "Emperors, government and bureaucracy," in A. Cameron and P. Garnsey (eds.) *The Cambridge Ancient History*. Vol. XIII *The Late Empire, AD 337–425* (Cambridge) 138–183

Kleberg, T. (1962) *Bokhandel och Bokförlag i Antiken*, Stockholm, Italian trans. E. Livrea as "Commercio librario ed editoria nel mondo antico," in G. Cavallo (ed.) *Libri, editori e pubblico nel mondo antico. Guida storica e critica* (Rome/Bari 1975) 13–83

Kleijwegt, M. (1991) *Ancient Youth. The Ambiguity of Youth and the Absence of Adolescence in Greco-Roman Society*, Amsterdam

Klein, J. (1934–36), *Die Griechische Logik und die Entstehung der Algebra*, Engl. trans. E. Brann, *Greek Mathematical Thought and the Origin of Algebra*, Cambridge, Mass./ London 1968

Klemm, F. (1954) *Technik: Eine Geschichte ihrer Probleme*, Freiburg/Munich

Knorr, W. R. (1978a) "Archimedes and the spirals," *Historia Mathematica* 5: 43–75

(1978b) "Archimedes' neusis construction in 'Spirals'", *Centaurus* 22: 77–98

(1978c) "Archimedes and the pre-Euclidean proportion theory," *Archives Internationales d' Histoire des Sciences* 28: 183–244

(1978d) "Archimedes and the Elements," *Archive for History of Exact Sciences* 19: 211–290

(1982a) "The hyperbola-construction in the 'Conics', book II: ancient variations on a theorem of Apollonius," *Centaurus* 25: 253–291

(1982b) "Observations on the early history of the Conics," *Centaurus* 26: 1–24

(1983) "Construction as existence proof in ancient geometry," *Ancient Philosophy* 3: 125–148

(1986a) *The Ancient Tradition of Geometric Problems*, Boston/Basel/Berlin

(1986b) "Archimedes' dimension of the circle: a view of the genesis of the extant text," *Archive for History of Exact Sciences* 35: 281–324

(1989) *Textual Studies in Ancient and Medieval Geometry*, Boston/Basel/Berlin

(1992) "When circles don't look like circles: an optical theorem in Euclid and Pappus," *Archive for History of Exact Sciences* 44: 287–329

(1993) "Arithmêtikê stoicheîosis: on Diophantus and Hero of Alexandria," *Historia Mathematica* 20: 180–192

Kordig, C. R. (1982) "The mathematics of mysticism: Plotinus and Proclus," in R. Baine Harris (ed.) *The Structure of Being. A Neoplatonic Approach* (Norfolk, Virg.) 114–121

Koyré, A. (1948a) "Les philosophes et la machine," in *Études d' Histoire de la Pensée Philosophique* (Paris 1971) 305–339

(1948b) "Du monde de l''à peu près' à l'univers de la précision," in *Etudes d'Histoire de la Pensée Philosophique* (Paris 1971)

Krafft, F. (1966) "Bemerkungen zur Mechanischen Technik," *Technikgeschichte* 33: 121–159

Lauffer, S. (ed.) (1971) *Diokletians Preisedikt*, Berlin

Levick, B. (1985) *The Government of the Roman Empire. A Sourcebook*, London/Sydney

Levin, F. R. (1975) *The Harmonics of Nicomachus and the Pythagorean Tradition*, The American Philological Association

Lewis, N. (1963) "The non-Scholars of the Alexandrian Museum," *Mnemosyne* 16.4: 257–261

(1965) "Exemption from liturgy in Roman Egypt," *The Bulletin of the American Society of Papyrologists* 2: 87–92

Lim, R. (1995) *Public Disputation, Power, and Social Order in Late Antiquity*, Berkeley/Los Angeles/London

Lizzi, R. (1990) "La memoria selettiva," in G. Cavallo, P. Fedeli and A. Giardina (eds.) *Lo spazio letterario di Roma antica* (Rome) 647–676

Lloyd, G. E. R. (1993) "Galen on Hellenistics and Hippocrateans: contemporary battles and past authorities," in J. Kollesch and D. Nickel (eds.) *Galen und das Hellenistische Erbe* (Stuttgart) 125–143

(1994) "Learning by numbers," *Extrême-Orient, Extrême-Occident* 16: 153–167

Lugli, G. (1970) *Itinerario di Roma antica*, Milan

Lynch, J. P. (1972) *Aristotle's School: A Study of a Greek Educational Institution*, Berkeley

MacMullen, R. (1964) "Social mobility and the Theodosian Code," *The Journal of Roman Studies* 54: 49–53

(1966) *Enemies of the Roman Order. Treason, Unrest and Alienation in the Empire*, London/New York

(1971) "Social history in astrology," in *Changes in the Roman Empire: Essays in the Ordinary* (Princeton/London 1991) 218–224

Mahoney, M. S. (1968) "Another look at Greek mathematical analysis," *Archive for History of Exact Sciences* 5: 319–348

Mansuelli, G. A. (1985) "Architetto e città," in M. Vegetti (ed.) *Il sapere degli antichi* (Turin) 180–200

Manuli, P. (1983) "Lo stile del commento. Galeno e la tradizione ippocratica," in F. Lasserre and P. Mudry (eds.) *Formes de pensée dans la collection hippocratique. Actes du IVe colloque international hippocratique, Lausanne 1981* (Geneva) 471–482

Marrou, H. I. (1963) "Synesius of Cyrene and Alexandrian NeoPlatonism," in A. Momigliano (ed.) *The Conflict between Paganism and Christianity in the Fourth Century* (Oxford) 126–150

(1965) *Histoire de l'éducation dans l'antiquité*, Paris

Marsden, E. W. (ed.) (1971) *Greek and Roman Artillery: Technical Treatises*, Oxford

Mazzarino, S. (1951) *Aspetti sociali del quarto secolo*, Rome

Menninger, K. (1958) *Zahlwort und Ziffer*, Göttingen, Engl. trans. P. Broneer, *Number Words and Number Symbols. A Cultural History of Numbers*, Cambridge, Mass./London 1969

Micheli, G. (1995) *Le origini del concetto di macchina*, Florence

Millar, F. (1977) *The Emperor in the Roman World (31 BC–AD 337)*, London

(1983) "Empire and city, Augustus to Julian: obligations, excuses and status," *The Journal of Roman Studies* 73: 76–96

Mitteis, L. (ed.) (1906) *Griechische Urkunden der Papyrussammlung zu Leipzig*, Leipzig

Mogenet, J. (1951) "La division selon Pappus d'Alexandrie," *Bulletin de la Classe des Lettres de l' Académie Royale de Belgique* 37: 16–23

(1956) "L'introduction à l'Almageste," *Mémoires de l' Académie Belgique* 51.2

(1961) "L'Histoire des isopérimètres chez les Grecs," *Scrinium Lovaniense: Mélanges Historiques Etienne Van Cauwenbergh*, special issue of *Recueil des Travaux d'Histoire et de Philologie* 4.24: 69–78

Momigliano, A. (1971) *The Development of Greek Biography*, Cambridge, Mass.

Molland, A. G. (1976) "Shifting the foundations: Descartes's transformation of ancient geometry," *Historia Mathematica* 3: 21–49

Morrow, G. R. (1970) *Proclus Commentary on the First Book of Euclid's Elements*, Princeton N.J.

Mueller, I. (1981) *Philosophy of Mathematics and Deductive Structure in Euclid's Elements*, Cambridge, Mass./London

(1982) "Aristotle and the quadrature of the circle," in N. Kretzmann (ed.) *Infinity and Continuity in Ancient and Medieval Thought* (Ithaca/London) 146–164

(1987a) "Mathematics and philosophy in Proclus' commentary on book I of Euclid's Elements," in J. Pépin and H. D. Saffrey (eds.) *Proclus. Lecteur et Interprète des Anciens* (Paris) 305–318

(1987b) "Iamblichus and Proclus' Euclid commentary," *Hermes* 115: 334–348

(1991) "On the notion of a mathematical starting point in Plato, Aristotle and Euclid," in A. C. Bowen (ed.) *Science and Philosophy in Classical Greece* (New York/London) 59–97

Mugler, C. (1958) *Dictionnaire Historique de la Terminologie Géométrique des Grecs*, Paris

Müller, A. (1910) "Studentenleben im 4 Jahrhundert n. Chr.," *Philologus* 69: 292–317

Müller, W. (1953) "Das isoperimetrische Problem in Altertum," *Sudhoffs Archiv* 37: 39–71

Napolitano Valditara, L. M. (1988) *Le idee, i numeri, l'ordine: la dottrina della 'mathesis universalis' dall'Accademia antica al Neoplatonismo*, Naples

Nesselman, G. H. F. (1842) *Die Algebra der Griechen*, Berlin

Netz, R. (1999) *The Shaping of Deduction in Greek Mathematics. A Study in Cognitive History*, Cambridge

(forthcoming a) "Greek mathematicians: a group picture," in C. Tuplin (ed.) *Science Matters*, Oxford

(forthcoming b) "The Aristotelian paragraph," *Proceedings of the Cambridge Philological Society*

(forthcoming c) *Archimedes. Works*, Engl. trans. R. Netz, Cambridge

Neuenschwander, E. (1974) "Zur Uberlieferung der Archytas-Lösung des delischen Problems," *Centaurus* 18: 1–5

Neugebauer, O. (1975) *A History of Ancient Mathematical Astronomy*, Berlin/New York/etc.

Neugebauer, O. & Van Hoesen, H. B. (1959) *Greek Horoscopes*, Philadelphia

Nixon, C. E. V. & Saylor Rodgers, B. (1994) *In Praise of Later Roman Emperors. The Panegyrici Latini*, Engl. trans. of the Latin texts R. A. B. Mynors, Berkeley/Los Angeles/Oxford

Nutton, V. (1971) "Two notes on immunities: Digest 27,1,6,10 and 11," *The Journal of Roman Studies* 61: 52–63

(1972) "Ammianus and Alexandria," *Clio Medica* 7: 165–176

(1977) "Archiatri and the medical profession in antiqui:y," *Papers of the British School at Rome* 45: 191–226

Oleson, J. P. (1984) *Greek and Roman Mechanical Wate⸗-Lifting Devices: The History of a Technology*, Toronto

O'Meara, D. (1989) *Pythagoras Revived: Mathematics and Philosophy in Late Antiquity*, Oxford

Panerai, M. C. & Filippi, M. R. (1984) "Come si fa la centuriazione," in *Misurare la terra: centuriazione e coloni nel mondo romano* (Modena) 109–150

Parroni, P. (1989) "Scienza e produzione letteraria," in G. Cavallo, P. Fedeli and A. Giardina (eds.) *Lo spazio letterario di Roma antica* (Rome) 468–505

Passalacqua, L. (1994) "Le *Collezioni* di Pappo: polemiche editoriali e circolazione di manoscritti nelle corrispondenze di Francesco Barozzi con il Duca di Urbino," *Bollettino di storia delle scienze matematiche* 14: 91–156

Pearcy, L. T. (1984) "Melancholy rhetoricians and melancholy rhetoric: 'black bile' as a rhetorical and medical term in the second century AD," *Journal of the History of Medicine and Allied Science.* 33: 446–456

Pease A. S. (ed.) (1967) *Aeneis IV*, Darmstadt

Pecere, O. (1990) "I meccanismi della tradizione testuale," in G. Cavallo, P. Fedeli and A. Giardina (eds.) *Lo spazio letterario di Roma antica* (Rome) 297–386

Pedersen, F. S. (1976) *Late Roman Public Professionalism,* Odense

Peirce, P. (1989) "The arch of Constantine: propaganda and ideology in late Roman art," *Art history* 12: 387–418

Pendlebury, R. (1873) "On a method of finding two mean proportionals," *The Messenger of Mathematics* 2.2: 166–168

Penella, R. J. (1990) *Greek Philosophers and Sophists in the Fourth Century AD. Studies in Eunapius of Sardis*, Leeds

Pingree, D. (1994) "The teaching of the 'Almagest' in late antiquity," in T. D. Barnes (ed.) *The Sciences in Greco-Roman Society*, special issue of *Apeiron* 27: 75–98

Pleket, H. W. (1973) "Technology in the Greco-Roman world: a general report," Ταλαντα 5: 6–47

Pricoco, S. (1985) "Filosofi e professori di filosofia nella tarda antichità: vecchi e nuovi modelli culturali fra IV e V secolo," in C. Giuffrida & M. Mazza (eds.) *Le trasformazioni della cultura nella tarda antichità* (Rome) 509–527

Regli, E. and Camaiora, R. (1984) "Che cos'è la centuriazione" in *Misurare la terra: centuriazione e coloni nel mondo romano* (Modena) 72–108

Rehder, W. (1982) "Die Analysis und Synthese bei Pappus," *Philosophia Naturalis* 19: 350–370

Reynolds, J. M. & Ward Perkins, J. B. (eds.) (n.d.) *The Inscriptions of Roman Tripolitania*, Rome

Reynolds, L. D. & Wilson, N. G. (1968) *Scribes and Scholars*, London

Rhys Bram, J. (1975) *Ancient Astrology: Theory and Practice*, Engl. trans. of Firmicus Maternus, *Matheseos Libri*, Park Ridge

Robinson, R. (1936) "Analysis in Greek geometry," *Mind* 45: 464–73

Rochberg, F. (1992) "The cultures of ancient science: some historical reflections – introduction," *Isis* 83: 547–553

Romano, E. (1987) *La capanna e il tempio. Vitruvio o dell'architettura*, Palermo

Romano, F. (1985) "Genesi e struttura del commentario neoplatonico," in C. Giuffrida and M. Mazza (eds.), *Le trasformazioni della cultura nella tarda antichità* (Rome) 219–237

    (1994) "La scuola filosofica e il commento," in G. Cambiano, L. Canfora and D. Lanza (eds.), *Lo spazio letterario della Grecia antica* (Rome) 587–611

Rome, A. (1938) "Un manuscrit de la bibliothèque de Boniface VIII à la Médicéenne de Florence," *L'antiquité classique* 261–270

    (1948) "Notes sur le ms. astronomique Norimbergensis gr. cent.V app. 8," *Scriptorium* 2: 113–117

Russell, D. A. (1989) "Arts and sciences in ancient education," *Greece and Rome* 36: 210–225

Sabra, A. I. (1987) "The appropriation and subsequent naturalizations of Greek science in medical Islam: a preliminary statement," *History of Science* 25: 223–243

Sachs, E. (1917) *Die fünf Platonische Körper*, Berlin

Saffrey, H. D. (1968) "ΑΓΕΩΜΗΤΡΗΤΟΣ ΜΗΔΕΙΣ ΕΙΣΙΤΩ. Une inscription légendaire," *Revue des études grecques* 81: 67–87

Saito, K. (1995) "Doubling the cube: a new interpretation of its significance for early Greek geometry," *Historia Mathematica* 22: 119–137

Saller, R. P. (1980) "Promotion and patronage in equestrian careers," *The Journal of Roman Studies* 70: 44–59

Salzman, M. R. (1990) *On Roman Time. The Codex-Calendar of 354 and the Rhythms of Urban Life in Late Antiquity*, Berkeley/Los Angeles/Oxford

de Santillana, G. (1961) *The Origins of Scientific Thought*, Chicago

Santini, C. (1990) "Le *Praefationes* dei gromatici," in C. Santini and N. Scivoletto (eds.) *Prefazioni, prologhi, proemi di opere tecnico-scientifiche latine* (Rome) 135–148

Sarton, G. (1938) "The tradition of Zenodorus," *Isis* 28: 461–462

Schemmel, F. (1909) "Die Hochschule von Alexandria im IV. und V. Jahrhundert p. Ch. n.," *Neue Jahrbücher für das klassische Altertum, Geschichte und Deutsche Literatur und für Pädagogik* 24: 438–457

Schindel, U. (1992) "Nachklassischer Unterricht im Spiegel der gromatischen Schriften," in O. Behrends and L. Capogrossi Colognesi (eds.) *Die Römische Feldmeßkunst. Interdisziplinäre Beiträge zu ihrer Bedeutung für die Zivilisationsgeschichte Roms* (Göttingen) 375–397

Schmidt, W. (1901) "Zur Geschichte der Isoperimetrie im Altertum," *Bibliotheca Mathematica* 2: 5–8

Schmitz, M. (1997) *Euklids Geometrie und ihre mathematiktheoretische Grundlegung in der neuplatonischen Philosophie des Proklos*, Würzburg

Schürmann, A. (1991) *Griechische Mechanik und antike Gesellschaft. Studien zur staatlichen Förderung einer technischen Wissenschaft*, Stuttgart

Seidenberg, A. (1966) "Some remarks on Nicomedes' duplication," *Archive for History of Exact Sciences* 3: 97–101

Sharples, R. W. (1990) "The school of Alexander?" in R. Sorabji (ed.) *Aristotle Transformed. The Ancient Commentators and their Influence* (London) 83–111

de Solla Price, D. J. (1974) "Gears from the Greeks. The Antikythera mechanism – a calendar computer from *ca.* 80 BC," *Transactions of the American Philosophical Society* 64: part 7

Solomon, J. (1983) "Vaticanus Gr.2338 and the εἰσαγωγή ἁρμονική," *Philologus* 127: 247–253

Stahl, W. H. (1964) "The systematic handbook in antiquity and the early Middle Ages," *Latomus* 23: 311–321

Straub, J. (1970) "Severus Alexander und die Mathematici," in J. Straub (ed.) *Bonner Historia-Augusta-Colloquium 1968/1969* (Bonn) 247–272

Szabó, A. (1974) "Analysis und Synthesis. Pappus II p.634 ff. Hultsch," *Acta Classica Universitatis Scientiarum Debreceniensis* 10–11: 155–164

Szabo, M. E. (1975) "Sporus of Nicaea," *DSB*

Tannery, P. (1880a) "L'arithmétique des Grecs dans Pappus," in *Mémoires Scientifiques I* (Toulouse/Paris 1912) 80–105

(1880b) "L'article de Suidas sur Hypatia," in *Mémoires Scientifiques I* (Toulouse/Paris 1912) 74–79

(1882) "Sur Sporos de Nicée," in *Mémoires Scientifiques I* (Toulouse/Paris 1912) 178–184

(1883) "Sur une critique ancienne d'une démonstration d'Archimède," in *Mémoires Scientifiques I* (Toulouse/Paris 1912) 300–316

(1883–4) "Pour l'histoire des lignes et surfaces courbés dans l'antiquité," in *Mémoires Scientifiques II* (Toulouse/Paris 1912) 1–47

(1884a) "Eutocius et ses contemporains," in *Mémoires Scientifiques II* (Toulouse/Paris 1912) 118–136

(1884b) "Domninos de Larissa," in *Mémoires Scientifiques II* (Toulouse/Paris 1912) 105–117

(1896) "Sur la religion des derniers mathématiciens de l'antiquité," in *Mémoires Scientifiques II* (Toulouse/Paris 1912) 527–539

Taub, L. C. (1993) *Ptolemy's Universe*, Chicago/La Salle, Ill.

Teitler, H. C. (1985) *Notarii and Exceptores. An Inquiry into the Role and*

Significance of Shorthand Writers in the Imperial and Ecclesiastical Bureaucracy of the Roman Empire (from the Early Principate to c. 450 AD), Amsterdam

Thaer, C. (1940) "Die Würfelverdoppelung des Apollonios," *Deutsche Mathematik* 5: 241–243

Thorndike, L. (1913) "A Roman astrologer as a historical source: Julius Firmicus Maternus," *Classical Philology* 8: 415–435

Tittel, C. (1910) "Geminos I," in *PW* 7.1: cols. 1026–1050

Tod, M. N. (1957) "Sidelights on Greek philosophers," *The Journal of Hellenic Studies* 77: 132–141

Tomei, M. A. (1982) "La tecnica nel tardo impero romano: le macchine da guerra," *Dialoghi di archeologia* 4: 63–88

Toneatto, L. (1992) "Il nuovo censimento dei manoscritti latini di agrimensura tradizione diretta e indiretta)," in O. Behrends and L. Capogrossi Colognesi (eds.) *Die Römische Feldmeßkunst. Interdisziplinäre Beiträge zu ihrer Bedeutung für die Zivilisationsgeschichte Roms* (Göttingen) 26–66

Toomer, G. J. (1972) "The mathematician Zenodorus," *Greek, Roman and Byzantine Studies* 13: 177–192

(1974) "Hypatia," *DSB*

(1976) see Diocles, *On Burning Mirrors*

(ed.) (1990) *Apollonius. Conics – Books V to VII*, New York/Berlin/etc.

Traina, G. (1994) *La tecnica in Grecia e a Roma*, Rome/Bari

Treweek, A. P. (1957) "Pappus of Alexandria. The manuscript tradition of the 'Collectio Mathematica'", *Scriptorium* 11: 195–233

Turner, A. J. (1994) *Mathematical Instruments in Antiquity and the Middle Ages: An Introduction*, London

Unguru, S. (1974) "Pappus in the thirteenth century in the Latin West," *Archive for History of Exact Sciences* 13: 307–324

van der Eijk, P. (1997) "Towards a rhetoric of ancient scientific discourse. Some formal characteristics of Greek medical and philosophical texts (Hippocratic Corpus, Aristotle)," in E. J. Bakker (ed.) *Grammar as Interpretation. Greek Literature in its Linguistic Contexts* (Leiden/New York/Cologne) 77–129

Vegetti, M. (1994) "Enciclopedia ed antienciclopedia: Galeno e Sesto Empirico," in G. Cambiano, L. Canfora and D. Lanza (eds.), *Lo spazio letterario della Grecia antica* (Rome) 333–359

Vera, D. (1981) *Commento storico alle Relationes di Quinto Aurelio Simmaco*, Pisa

Ver Eecke, P. (1933a) see Pappus Alexandrinus, *La Collection mathématique*

(1933b) "La mécanique des Grecs d' après Pappus d'Alexandrie," *Scientia* 54: 114–121

(1948) *Proclus de Lycie. Les Commentaires sur le premier livre des Eléments d' Euclide*, French trans., Bruges

(1956) "Eutocius et sa tradition de la lettre d'Eratosthène au roi Ptolémée sur la duplication des cubes," *Archives d'histoire des sciences* 9.35: 217–226

Vernant, J.-P. (1965) *Mythe et pensée chez les Grecs. Etudes de psychologie historique*, Paris

Visky, K. (1959) "La qualifica della medicina e dell'architetura nelle fonti del diritto romano," *Iura* 10: 24–66

Vitrac, B. (1992) "A propos de la chronologie des oeuvres d'Archimède," in J.-Y. Guillaumin (ed.) *Mathématiques dans l'antiquité* (Saint-Étienne) 59–93

— (1994) "Euclide et Héron: Deux approches de l'enseignement des mathématiques dans l'Antiquité?" in G. Argoud (ed.) *Science et vie intellectuelle à Alexandrie* (Saint-Étienne) 121–145

— (1996) "Mythes (et réalités?) dans l'histoire des mathématiques grecques anciennes," in C. Goldstein, J. Gray and J. Ritter (eds.) *L'Europe mathématique. Histoire, mythes, identité* (Paris) 33–51

van der Waerden, B. L. (1954) *Science Awakening*, Engl. trans. A. Dresden, Groningen

Wantzel, P. L. (1837) "Recherches sur les moyens de reconnaître si un problème de géométrie peut se résoudre par la règle et le compas," *Journal de Mathématiques* 2: 366–372

Waterhouse, W. C. (1972) "The discovery of the regular solids," *Archive for History of Exact Sciences* 9: 212–221

Westerink, L. G. (1961) "Elias on the Prior Analytics," *Mnemosyne* 14: 126–139

Wiedemann, T. (1979) "Petitioning a fourth-century emperor: the *De Rebus Bellicis*," *Florilegium* 1: 140–150

Willem, V. (1928–30) "L'Architecture des abeilles," *Bulletin de la classe des sciences de l'Académie Royale de Belgique* 14–16

Williams, B. P. & Williams, R. S. (1995) "Finger numbers in the Greco-Roman world and the early Middle Ages," *Isis* 86: 587–608

Williams, S. (1985) *Diocletian and the Roman Recovery*, London

Wilson, N. G. (1984) "The relation of text and commentary in Greek books," in C. Questa and R. Raffaelli (eds.), *Atti del convegno internazionale 'Il libro e il testo' Urbino 1982* (Urbino) 104–110

Wirth, G. (1980) "Vom Anonymus de Rebus Bellicis zu Ammian. Perspektiven eines Rombildes im 4. Jhdt.," in *Passagio dal mondo antico al medioevo. Da Teodosio a san Gregorio Magno* (Rome) 87–122

Zanker, P. (1995) *The Mask of Socrates. The Image of the Intellectual in Antiquity*, Engl. trans. A. Shapiro of *Maske des Sokrates*, Berkeley/Los Angeles/Oxford

Zeuthen, H. (1896) "Die geometrische Construction als 'Existenz-beweis' in der antiken Geometrie," *Mathematische Annalen* 47: 222–228

Zhmud, L. (1997) *Wissenschaft, Philosophie und Religion im frühen Pythagoreismus*, Berlin (*non vidi*)

Ziegler, K. (1896) "Pappos 2," in *PW* 18: cols. 1084–1106

GENERAL INDEX

accountants and calculation, *see* mathematical professions
accuracy, 13, 24, 120, 189
administration and bureaucracy, 21 f., 26 ff., 40 f.
Aelian, 36n101, 88
Aelius Aristides, 36n103, 51
Agennius Urbicus, 24 f.
Alexandria, 5, 32, 43 ff., 50, 52, 85, 96, 99, 137, 196
Ammianus Marcellinus, 11n3, 45, 101, 124n70
Apollodorus, 125
Apollonius, 60n6, 79 f., 85, 118, 128n3, 136, 139, 147, 151 f., 154 f., 164n96, 181 ff., 187 ff., 196 ff.
Aratus, 46
Archimedes, 9, 19, 23, 59 f., 62 ff., 70 ff., 77, 79 f., 82, 86, 91 ff., 100 ff., 110, 113, 117, 122, 128n3, 141, 143n46, 151 f., 158 ff., 163 f., 174 ff., 187 ff., 196 ff.
architecture and architects, *see* mathematical professions
Archytas, 55, 83, 91, 134, 136n33, 137
Aristaeus, 60n6, 61n12, 136, 188n45, 196
Aristarchus, 73, 142n43
Aristotle, 12, 38n113, 43, 50, 52, 54, 71, 77, 80, 82, 104 f., 155, 163, 172 ff., 178; pseudo-Aristotle, 97 ff., 117
astrology and astrologers, *see* mathematical professions
astronomy and astronomers, *see* mathematical professions
Athenaeus, 101n20
Athens, 43n131, 51
Augustine, 19 f., 27n71, 43, 46, 48 f.
Autolycus, 67n39, 73

Babylonians, 20, 42
Balbus, 23 f.
Biton, 101n18, 125

calculatores, see accountants
Carpus of Antioch, 108, 117, 171 f.
Cassiodorus, 19, 124
Chaldeans, *see* Babylonians
Cicero, 66, 72n50, 102n22
Claudian, 102
commentaries, 29, 53, 68, 85, 147, 163, 169
Constantine, Arch of, 1, 169
Constantinople, 42, 43

Del Monte, Guidobaldo, 7, 113n48
Demetrius Poliorketes, 100 f.
De rebus bellicis, 33n92, 123 ff.
Descartes, 7
Dio Cassius, 36n103
Diocles, 136n31, 147 f., 152
Diocletian's Price Edict, 30, 38
Diodorus, 42n129, 101n18, 101n19
Dionysius of Syracuse, 101
divination and diviners, 14, 15, 19, 37 f., 45, 49
Dorotheus Sidonius, 11

education, *see* teaching
Egypt and Egyptians, 14n18, 20, 42, 45, 121, 125
Emperor, Roman, 11, 20, 21n44, 23, 26, 29, 30, 32, 33, 34, 36, 37 f., 40 ff., 48, 101, 123 ff.
engineering and engineers, *see* mechanics
Epicureans, 83
Eratosthenes, 5n8, 55, 83, 134 ff., 142, 147, 150, 161
ethics, 11, 32, 35, 37 ff., 40, 52, 54, 129, 133, 170, 193 ff.
Euclid, 19, 46, 55, 59 f., 62n18, 66, 69 ff., 73 f., 76n61, 78 ff., 86, 139, 154, 156, 174, 179n24, 180n25, 196 ff.
Eudemus, 85, 136n33
Eudoxus, 55, 66, 91, 134, 137
Eunapius, 21n44, 32n89, 36n101, 44, 45, 84n82, 106n32

224

INDEX LOCORUM